Locomotives and Rolling Stock of the London, Tilbury and Southend Railway

by

R.W. Rush

THE OAKWOOD PRESS

ISBN 0 85361 466 0

Typeset by Oakwood Graphics.

Printed by Alpha Print (Oxford) Ltd, Witney, Oxon.

Plate 1: At the front of a line of locomotives is 4-6-4T No. 2196, ex-London Tilbury & Southend No. 91. Built by Beyer, Peacock & Co. of Manchester in 1912, it carries the LMS number of 1929, having been No. 2104 in 1912. This locomotive was finally withdrawn in December 1932. *H.C. Casserley*

Other Oakwood Press books by the same author:
- East Lancashire Railway
- Furness Railway: Locomotives and Rolling Stock
- Lancashire & Yorkshire Passenger Stock
- North Staffordshire Railway Locomotives and Rolling Stock

Published by
The OAKWOOD PRESS
P.O. Box 122, Headington, Oxford OX3 8LU

Contents

Plate 2: '51' class 4-4-2T built in 1900 by Sharp, Stewart & Co., Glasgow (makers No. SS 4663) seen here carrying its 1912 number of 2168, later to become LMS 2102. This locomotive was withdrawn in August 1949. *H.C. Casserley*

Plate 3: '51' class 4-4-2T No. 2165 seen here passing Upminster with a 13 coach train on 22nd August, 1926. H.C. Casserley

Author's Note

The London, Tilbury & Southend Railway (or 'Tilbury' as it was usually known) was an almost unique railway, the nearest comparable company, perhaps, being its close neighbour the North London Railway, with which it had much in common. Both companies derived their chief income from passenger traffic, unlike most larger companies, whose main source of revenue was freight. Goods traffic on the Tilbury line was a secondary concern, the commuter service to and from Southend being the mainstay of the company's finances. With a total length of under 80 miles, and a locomotive stock not quite reaching 100, it was a small concern, but its importance far exceeded its size. It was inevitable that sooner or later it would be swallowed up by some larger company, and it was no surprise when in 1912 the Midland Railway absorbed it. To give the Midland its due, it very astutely let the Tilbury section continue largely on its own sweet way, only transferring heavy locomotive repairs to Derby, and it was not until the LMS group had been formed in 1923 that any major changes in working practices and services became apparent.

This is not intended to be a detailed history of the LT&SR, though a certain amount of history must be introduced to set the scene. It has long been the author's opinion that the LT&SR would make an excellent prototype for a model railway, and it is chiefly with modelling in mind that this book has been prepared. Hence the line drawings of most of the rolling stock, prepared mainly from photographs and published dimensions. Luckily a good deal of detail as to numbering, painting, etc. has been handed down, and photographs are fairly plentiful on the whole, particularly of locomotives, but less so with regard to rolling stock. A few details may be contentious to a degree, since the company's official diagrams were not very explicit, and on the whole were rather crude. The author must express his grateful thanks to Mr R.J. Essery's two-volume work on the Midland Railway goods wagons, which has been of great assistance in producing this book, and can be thoroughly recommended to all model railway enthusiasts, as well as to historians. My most grateful thanks are also due to Mr R.W. Kidner for the most helpful suggestions and criticisms he has made to this treatise.

ROBERT W. RUSH
Accrington

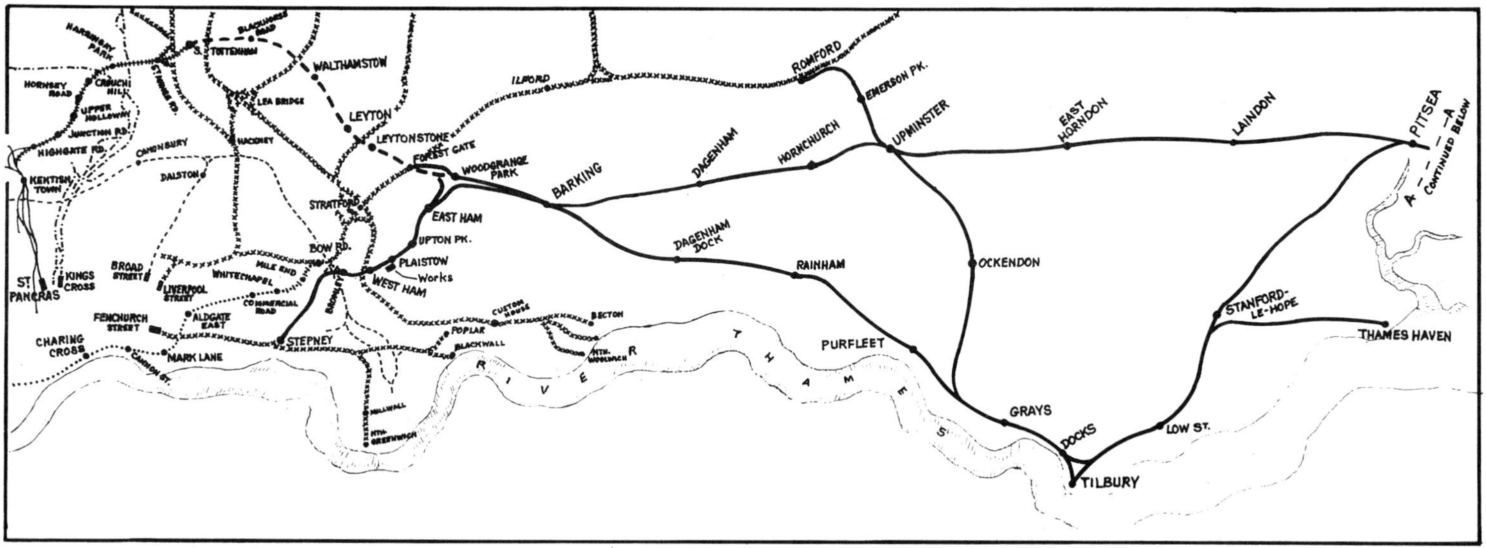

LONDON, TILBURY & SOUTHEND RAILWAY AND ITS CONNECTIONS, 1912.

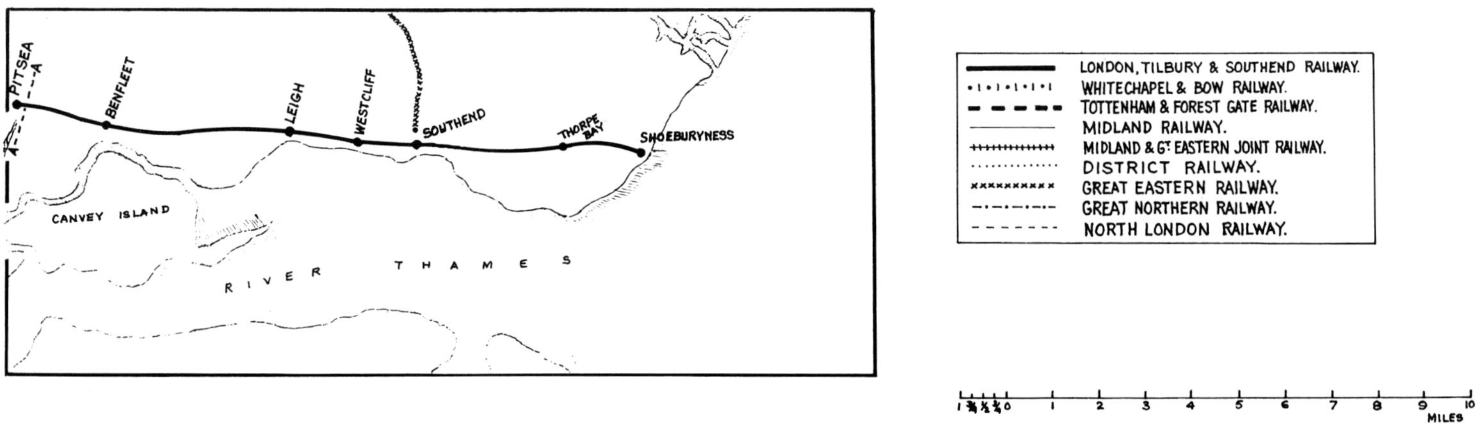

Chapter One

Historical Notes

The LT&SR was in a rather unusual position, in that for the the first twenty years or so of its existence it possessed no locomotives or rolling stock of its own, and was operated by another company on behalf of the two companies which jointly owned it. After this period it became fully independent, and ran the system itself for the rest of its career. This state of affairs came about from the manner of its inception. It was proposed by the London & Blackwall and Eastern Counties Railways (ECR) jointly, by a Bill passed on 17 June 1852 for a line commencing at a junction with the Eastern Counties Colchester route half a mile beyond Forest Gate station, and proceeding via Barking, Rainham and Purfleet to Tilbury, where it curved off through the back of the town and continued onwards to Stanford-le-Hope and Southend. At Tilbury two short branches, one from either side of the curve, joined together at the waterfront. It was a somewhat circuitous route to Southend, but Tilbury was the main objective, with its ferry service across the Thames to Gravesend in Kent and the pleasure gardens nearby in mind.

The working of the line was contracted out to Peto, Brassey, & Betts, the contractors who built it, for twenty-one years from the opening date, which turned out to be 3rd July 1854. The contractors were bound by the agreement to pay a dividend of 6 per cent on the share capital of £400,000. This agreement was not entirely satisfactory to the ECR shareholders, who had suspicions as to how the contract had been drawn up. The joint managing committee of three Directors from each of the two companies had pledged the ECR to find the capital for the rolling stock of the LT&SR at a very unfavourable rate; the ECR had paid £46,000 in the first year of operation for rolling stock and had only received £4,000 in return, or rather less than 1 per cent on its outlay. Nevertheless, nothing could be done, and the shareholders had to endure the conditions, though not without a good deal of recrimination. The two railways which Jointly proposed the LT&SR were entirely separate companies, though in the course of time they did come under the same banner, for in 1862 the Eastern Counties amalgamated with several other small railways to form the Great Eastern Railway (GER), and four years later the Blackwall Railway was leased to the GER.

The first section of the new line, from Forest Gate to Tilbury, was opened for traffic on 13 April 1854. The Bishopsgate terminus of the ECR was used for the Tilbury trains, and two days later Tilbury trains also began using the Fenchurch Street terminus of the Blackwall Railway. Trains from each terminus were combined at Stratford. The line was double throughout, with stations at Barking, Rainham, Purfleet, Grays, and Tilbury Fort. During the latter half of 1854 the line was extended from Tilbury to Stanford; and on 1st July, 1855 was complete from Tilbury to Southend; originally this part was single track. No telegraph system was installed, and as the Board of Trade declined to allow passing loops without telegraphic control, the Tilbury-Southend section had to be worked by one engine in steam. This allowed of only three trains each way daily, and as the line obviously could not be worked profitably under these conditions, orders were given to double the line forthwith, the necessary capital being furnished by the lessees.

An Act was passed on 7th July, 1856 for a line from Gas Factory Junction (Whitechapel) on the Blackwall Railway via Bromley, Plaistow and East Ham, to Barking. This line would help to avoid the delays caused by the heavy traffic through Stratford, and it was opened on 31st March, 1858. From then onwards Tilbury trains ceased to use Bishopsgate station, all starting from, or terminating at, Fenchurch Street. The ECR substituted a local service between Bishopsgate and Barking, via Stratford. The Tilbury company was granted perpetual running powers over the Blackwall Railway from Fenchurch Street to Gas Factory Junction (2¾ miles) while in return the ECR received running powers into Barking.

At a special meeting of shareholders on 17th December, 1861, it was proposed that the LT&SR should be reconstituted into an entirely separate company, independent of either the Blackwall or Eastern Counties Railways, with its own directorate, capital, and share stock. The proposal was bitterly opposed by the Chairman of the ECR (who was one of the sitting directors of the LT&SR) but nevertheless it was passed by an overwhelming majority of the shareholders. The Bill was put before Parliament and was approved on 16th May, 1862. One of the clauses of this Bill laid down that the directorate should consist of three members elected by each of the two companies, plus three nominated by the shareholders of the LT&SR, these three to be completely independent of the other two companies. This gave rise to a considerable amount of dissatisfaction, since considering the Eastern Counties and Blackwall Railways still nominated three directors each, the Board as a whole could not be said to be 'independent'. About this time too, the Great Eastern Railway was constituted, the dominant partner being the ECR, but the Blackwall Railway did not come into the fold at that time - not for another four years.

Peto, Brassey & Betts petitioned for an alteration in the terms of the lease, whereby the ruling 6 per cent dividend should be reduced to 4½ per cent and the lease transferred lock, stock and barrel to the GER. This the LT&SR shareholders refused to countenance, and they set up a commission to look into the whole matter of the lease and its conditions. Their report was issued in February 1864, in which they stated that the 1862 Act had been promoted and entirely funded by Peto, Brassey & Betts with one object only - that of bringing forward the subsequent proposal for modification of the lease. The commission also uncovered the fact that the three LT&SR Directors were aware in advance of these machinations, and had connived at them. The commission's comments on the tolls were succinct and to the point; they alleged that there had been a considerable amount of skullduggery in drawing up the conditions. Prior to the construction of the Bow-Barking line the tolls fixed for running over the appropriate sections of the ECR and Blackwall Railway were one third of the through fare to Gravesend (including the ferry) a relevant proportion from intermediate stations, and one fourth of the fare to stations beyond Tilbury. There was general agreement that these tolls were extortionate. The construction of the Bow-Barking cut-off rendered it unnecessary for Tilbury trains to run over any ECR metals, the only running powers exercised in full being over the 2¾ miles of the Blackwall Railway to Gas Factory Junction. This made it unnecessary to pay the ECR anything so in order to secure against any opposition from the ECR to the new line, Peto, Brassey & Betts had agreed to pay the latter the same tolls as would have been necessary had the new line not been built. The Tilbury shareholders were incensed against the lessees, and demanded that the lease, be revoked forthwith, but it was pointed out that the lease had still 11 years to run, and that legally there was nothing that could be done, short of setting the whole affair before Parliament which the shareholders were loth to do. However, they demanded that tolls cease to be paid to the GER for the lines over which the Tilbury company did not run.

The second big stumbling block was the representation on the Board. In their report the commission stated 'the three shareholders' Directors each retire periodically, as in most other companies, and are eligible for re-election, but the six Blackwall and Eastern Counties Directors are not elected by you, but are sent by the two companies; you cannot remove them, and your obligation to receive them does not terminate with the lease, but exists for all time . . . Is it necessary to say that the sole purpose for which they sit on your Board is solely to protect Blackwall and ECR, interests?'

In conclusion, the report described the future of the line as very satisfactory, envisaging the rapid growth of Southend as a holiday resort, the construction of new docks at Dagenham and Tilbury; the land between Bow and Barking had still not been much built upon, but this would come to pass soon. All these prophecies were realised within a comparatively short time. As a result of the report, a Mr Eley was elected to the Board in March 1864, followed by Mr Moxon next year, both in replacement of retiring LT&SR Directors, Mr Eley becoming Chairman in 1865 in place of Mr Daniell of the Blackwall Railway, who had resigned. Mr Moxon unfortunately died in 1869, and was replaced by Mr Stockdale, who along with Messrs Eley and Moxon had always been staunchly on the side of the LT&SR. Stockdale and

Eley both retired in 1875, when the lease expired.

From 1867 onwards there was continual dissatisfaction with the Lessees, Peto, Brassey & Betts, who were accused of making no attempt to develop traffic and promote the facilities of the railway. They hid behind the statement that the expense necessary to do so could not be found because of the extortionate tolls they had to pay, though it was pointed out to them that such a situation was entirely their own fault in promoting the 1862 Act, and the remedy lay in their hands alone. Once again the shareholders found themselves powerless to act in their own interests.

The lease to Peto, Brassey & Betts was due to expire on 3rd July, 1875 and in preparation for taking over the management and operation themselves, the LT&SR Board appointed a Mr Underwood, of the Manchester, Sheffield & Lincolnshire Railway as adviser to the company. Discussions were then opened with the GER with a view to the latter taking over the entire concern, but this the GER declined to do, though they did agree to continue to supply rolling stock until 1878, and locomotives until 1880, to work the line temporarily until the LT&SR were in a position to obtain their own stock. The company then made arrangements for the purchase of some 250 wagons and a number of passenger coaches, all of which were delivered by late 1878. The purchase of locomotives was not so easy, as various types had to be examined and builders consulted, as to which would be the best for the purpose. Eventually twelve 4-4-2 tank locomotives were delivered in 1880 by Sharp, Stewart & Co. of Manchester, these forming the basis of most subsequent locomotive stock. They will be dealt with fully in the next section of this book. An Act was passed on 13th May, 1875 enabling the LT&SR to enter into an agreement to lease or sell the undertaking to any other railway company who might be interested, also to raise additional capital for the purpose of purchasing rolling stock, etc., and to renew the powers for operating the ferry service to Gravesend. As it turned out, the first section of this Act was unnecessary, since the company decided to retain full control of the system.

On 1st July, 1875 the company appointed its own Engineer and General Manager, in the person of Arthur Lewis Stride, who had been previously with the London, Chatham & Dover Railway. He was destined to hold this position (as well as that of Chairman) until the end of the company's independent existence, 37 years later. As he said himself at the last half-yearly meeting of the LT&SR in 1912, 'when we took over the line there was no rolling stock of any description, there was no telegraph system, there were no block signals, or anything of that kind'. The directorate decided that it was imperative that the line should be brought up to first class standards as soon as possible, when the necessary additional capital was available. The telegraph and the block system were installed gradually throughout, and further rolling stock and locomotives ordered. The agreement wlth the GER expired on 1st July, 1880, and by that time the LT&SR had twelve locomotives and sufficient rolling stock to maintain a reasonable service. Mr Thomas Whitelegg was appointed as locomotive superintendent late in 1879, which post he held until 1910, when he was succeeded by his son, Robert Harben Whitelegg; the latter remained at Plaistow after the Midland take-over until 1918, when he left to take up a similar post on the Glasgow & South Western Railway at Kilmarnock. Locomotive, carriage and wagon works were established at Plaistow (West Ham), and all kinds of work were undertaken there; though no new locomotives were actually built at Plaistow, a good deal of rebuilding was carried out.

A Bill was drafted in 1877 for the extension of the railway a further five miles from Southend to Shoeburyness, where there was a considerable Army establishment. Unfortunately it had to be dropped, owing to various objections from the War Office, but the Government decision was reversed in 1882, and the proposed extension to Shoeburyness Fort was authorised by an Act dated 24th July, 1882 and opened on 1st February, 1884. By this same Act, a new cut-off line was authorised to run direct from Barking via Dagenham, Upminster, and Laindon to Pitsea, cutting the overall distance to Southend by six miles. The same Act also repealed the 1862 Act, by which the GER and Blackwall Railway were empowered to appoint three directors each to the LT&SR Board. At long last the company got rid of this incubus which had stifled their progress for so many years. The Board was now reduced to seven members, all of whom must be shareholders with at least £1000 of stock. The new directorate was elected on 1st August, 1882, with Arthur Stride as Chairman. The LT&SR now became a completely independent company, controlled by its own shareholders, though the GER and the Blackwall Railway continued to have a minority shareholding.

The East & West India Dock Company, who owned a considerable part of the dock areas in the Port of London, were experiencing a lack of trade, and it was clear that something would have to be done to alleviate the congestion in the River. This company proposed to build a new series of docks at Tilbury in an effort to reduce the amount of large shipping going up the Thames, and also to try and attract more trade to the lower reaches of the River. For this proposal, the LT&SR gave its wholehearted backing, and a Bill for the construction of Tilbury Docks was approved on 3rd July, 1882. Work began forthwith. The cost was estimated at a million pounds, but in the event turned out to be three times this figure. It was agreed that the railway should build a new station at Tilbury Docks with all the best facilities (this was opened in 1885), and provide an hourly service to and from Fenchurch Street. A new warehouse was also built at Commercial Road (Whitechapel) for the use of the dock company - who took it over entirely in 1887.

The first eight miles of the Barking-Pitsea line, as far as Upminster, were put into traffic on 1st May, 1885, followed by the next four miles to Horndon a year later, but the final section over the Laindon Hills to Pitsea was not completed until 1st June, 1888. From this date, all the principal trains to Southend were diverted over the new line, though some did continue to run through via Tilbury.

As it happened, the new docks were very slow to attract shipping, and only by greatly reduced charges was any traffic persuaded to use them. By 1888 the dock company was in the hands of the official receiver, and though trade did pick up gradually it was not until the passing of the London & India Docks Amalgamation Act in 1900, and the Port of London Act of 1908, that the chaotic situation in London's docks was brought into some semblance of order. Both these Acts included clauses protecting the interests of the LT&SR. Today, a century later, Tilbury provides the main dock facilities for the Port of London, and very few of the larger steamers go up river.

A further extension of the railway was proposed in 1883, by the projection of a line from Romford on the GER main line to East Anglia, through Upminster to Grays, on the Tilbury loop. This was promulgated to forestall a threatened attempt by the GER to invade LT&SR territory and gain access to Tilbury Docks. The Act was passed on 20th August, 1883, but as the shareholders saw little real advantage in it, construction was delayed, and after extensions of time had been granted in 1886 and 1888, the line was eventually opened on 1st July, 1892, single track throughout, with only one intermediate station (Ockendon) between Upminster and Grays. The final section, from Upminster to Romford, did not open until 7th June, 1893.

The Tottenham & Forest Gate Railway was proposed in 1889 as a joint venture of the Midland and LT&SR, and the Bill for its construction was approved on 4th August, 1890, the line to commence at a junction with the Midland & Great Eastern Joint line at South Tottenham, and extend for six miles to join the LT&SR at Forest Gate, with a connecting curve to allow direct access to East Ham. Construction through the northern suburbs was somewhat difficult, including a substantial viaduct in brick, but the line was ready for traffic by 9th July, 1894. The Midland then began running local services between St Pancras and East Ham, a new bay platform having been constructed specially at the latter station. The LT&SR also began running trains through between St Pancras and Southend, two locomotives being stabled in the Midland shed at Kentish Town for this purpose. The Midland also began running into Tilbury Docks.

In 1892 began a period of intense competition with the GER for the Southend traffic. The GER had its own line to Southend, via Stratford, Romford, Billericay and Rochford, entering Southend from the north; it was some five miles longer than the LT&SR direct route via Laindon.

Plate 4: '37' class 4-4-2T No. 2138 seen here at Barking on 12th January, 1929. The locomotive was built by Sharp, Stewart & Co. Glasgow (makers No. SS 4248) in 1897 and entered the fleet as No. 40 *Black Horse Road*. It was subsequently rebuilt in 1911 and 1920. The locomotive was renumbered as No. 2149 in 1912, then No. 2138 in 1929 and finally No. 1956 in 1946, it was withdrawn in May 1951. *H.C. Casserley*

Plate 5: 'No. 1' class 4-4-2T No. 2132 built by Sharp, Stewart & Co. in 1884 and seen here at Bow shed in August 1925. Finally withdrawn in 1930. *H.C. Casserley*

Nevertheless, competition was spirited, especially as the City termini of the two railways were virtually next door to each other. In 1892 also, the GER had abolished second class, and the LT&SR felt obliged to follow suit in 1893. There were 13 GER trains daily from Liverpool Street to Southend, and in 1892 it put on 12 more; to counter this the LT&SR increased its service from Fenchurch Street from 10 to 16 trains daily. By 1895 there were seven from St Pancras.

As far back as 1882 a proposal had been made for a direct connection between the LT&SR and the Metropolitan District Railway in the region of Whitechapel, where the District's eastern line ended, but it came to nothing. It was not until August 1897 that an Act was passed for the construction of the Whitechapel & Bow Railway (W&B), a line two miles in length from the District's Whitechapel terminus to Campbell Road Junction, a short distance from where the LT&SR left the GER at Gas Factory Junction. This was to be the joint property of the District and LT&S Railways, with the former granted running powers to East Ham. In spite of its short length, it was difficult to build, being in a shallow tunnel beneath the Mile End Road for most of its length, only emerging at Bow Road to join the Tilbury line by a short sharp rise of 1 in 45. This line relieved Fenchurch Street of an amount of short distance traffic which now passed directly onto the District Railway via Mark Lane to Aldwych or Mansion House. Metropolitan District Railway steam trains started running over the W&B Railway on 2nd June, 1902, and went as far as East Ham; Metropolitan District Railway engines were assisted by four modified class '1' LT&SR engines for stand-by and special duties. This led to congestion on the track, and powers were obtained to quadruple the Tilbury line between Campbell Road and Barking. The work was begun immediately, and carried out with expedition, though it involved some large alterations, including the replacement and lengthening of the girder bridge over the North London Railway and sidings at Bow. The widening as far as East Ham was completed in 1905, but onwards to Barking took more than another two years, since a great deal of modifications were necessary, including realignment of the track, removal of a level crossing, and the rebuilding of Barking station, which had to be enlarged.

Meanwhile negotiations were under way for the electrification of the entire Metropolitan District Railway system, and in connection with this the Tilbury railway obtained powers in 1902 for the electrification of its own lines between Whitechapel and Barking, and arrangements were made to draw power from the London Underground Railways generating plant at Lots Road. Electric trains began running as far as East Ham in 1905, but were not extended to Barking until April 1908, when all the modifications had been completed. The LT&SR acquired a half share in 44 electric motor cars and four trailers of District Railway design as its share of operations.

There was one branch line which has not been mentioned so far. In fact, it rather tended to be forgotten through most of its career, though strangely enough at one period it became quite important. The branch began life as an independent company in 1836, as the Thames Haven Dock & Railway Co., with powers to construct a line from Romford to Thames Haven, but nothing was done. After reconstitution of the company in 1842 again nothing tangible was achieved. The company remained in being, however, for in September 1852 they approached the LT&SR (then under construction) to build a single line branch four miles long from a junction with the Tilbury-Southend line near Mucking, just south of Stanford, to Thames Haven, and abandon the rest of the line to Romford. It was also agreed that the Thames Haven company should construct wharves at the latter place, up to a cost of £10,000. The LT&SR agreed to these proposals and construction of the line went ahead; it was opened for traffic, worked by the lessees, on 7th June, 1855. Later that same year the LT&SR agreed to purchase the branch, the wharves, and twenty acres of land for £48,000 of LT&SR stock.

The object of the branch was to provide a port for traffic from British coastal resorts, such as Margate, Ramsgate and Sheerness, also round the Essex and Kent coasts as far as Ipswich and Dover. It was also hoped that it would become a landing place for fish traffic to the City. Two express trains were scheduled daily from Fenchurch Street to Thames Haven, via Tilbury, at 10.22 am and 4.07 pm, booked to connect with the boats of the General Steam Navigation Co. to Margate. In the opposite direction the trains left Thames Haven after the arrival of the 10.20 am and 4 pm boats from Margate, the train journey taking approximately 56 minutes. During 1856 it was reported that the service was very successful, aided no doubt by the very cheap fares charged on the boats. From 1869 to 1886 through trains were also run from stations in North London.

Agreements were made in 1865 to import cattle through Thames Haven from France and Holland, but this traffic did not start until 1867, due to an outbreak of foot and mouth disease. Thereafter the trade went on by fits and starts until 1892, broken by various outbreaks of cattle diseases, the Franco-Prussian War of 1870, and on one occasion at least, by the intervention of the Privy Council - though it is not stated why that august body took such drastic action. A proposal was made that all cattle slaughtering should be done at Thames Haven, but this was vetoed by the London butchers, who strongly objected to having to travel out to Thames Haven to buy their meat. So most of the cattle were shipped by the LT&SR to Bromley, where they were handed over to the North London Railway for transfer to the slaughterhouse at Maiden Lane. Equipment at Thames Haven was very good, with accommodation for up to 2,000 cattle and 8,000 sheep. In these comparatively early days the port was very successful, but with the gradual rise of Tilbury Docks, which were even better equipped, from 1886 to 1896 the trade at Thames Haven declined, and by 1912 passenger traffic had ceased. A bit of a revival occurred during World War II, but traffic generally had sunk into a state of somnolence, only being kept from dying out entirely by the oil refinery at Shell Haven and the Kynoch works near Corringham.

On 1st June, 1901 the branch had been connected near its end with the Corringham Light Railway, built to carry workers from Corrringham village to an explosive works, then called Kynochtown. From about 1915 three ex-LT&SR four-wheeled coaches were used, (Nos. 2, 7, 9 - at that time renumbered Midland Railway 02445/49/51) and during World War I some old Midland Railway coaches were also used. The Kynoch factory was purchased by Messrs Cory, and in 1921 the station at Kynochtown was renamed Coryton.

By the latter part of 1911 it became apparent that the LT&SR could not survive much longer independently, even though it was in a strong financial position and was an efficient and compact railway. As a small company, with only about 60 route miles, and some 80 locomotives, situated in an intensely worked area, the writing was on the wall. This had been demonstrated in 1908, when the only other comparable railway in the area, the North London, had gone into full collaboration with the London & North Western, a situation only just short of complete amalgamation. There were really only two possible large companies to take over the Tilbury line, these being the Great Eastern and the Midland, though the LNWR did display some interest, but the Midland made the first overtures, and eventually the best offer. At first the GER strongly opposed the merger, but later withdrew, and agreed to enlarge Fenchurch Street station; also to equip its own section of the Blackwall Railway (Fenchurch Street to Gas Factory Junction) for electric working, since the Midland had expressed its intention of electrifying the whole of the LT&SR system. However, the outbreak of war in 1914 effectively put a stop to this proposal, and it was never carried out. Arthur Stride, who had been the Tilbury company's General Manager and Chairman for the whole of its independent existence, was then 70 years of age, and not in the best of health, so doubtless this had some bearing on the decision to sell out. The Midland offered a very good price, and the shareholders received £240 of Midland 2½ per cent preference stock for every £100 of Tilbury stock, so they were well satisfied with the deal. The Act for the amalgamation was passed on 7th August, 1912. By this Act the GER was granted running powers over the Tottenham & Forest Gate Railway into St Pancras, and the GNR was given running powers to Tilbury Docks.

So passed an efficient, well managed small railway company which had an excellent record of safety and punctuality, carrying an enormous number of commuters to and from the City every day from all points as far as Southend.

The morning and evening commuter expresses were very well patronised. To give the Midland Railway their due, they continued the good work right through World War I amid countless difficulties, until they themselves disappeared in the main Grouping of Railways in 1923.

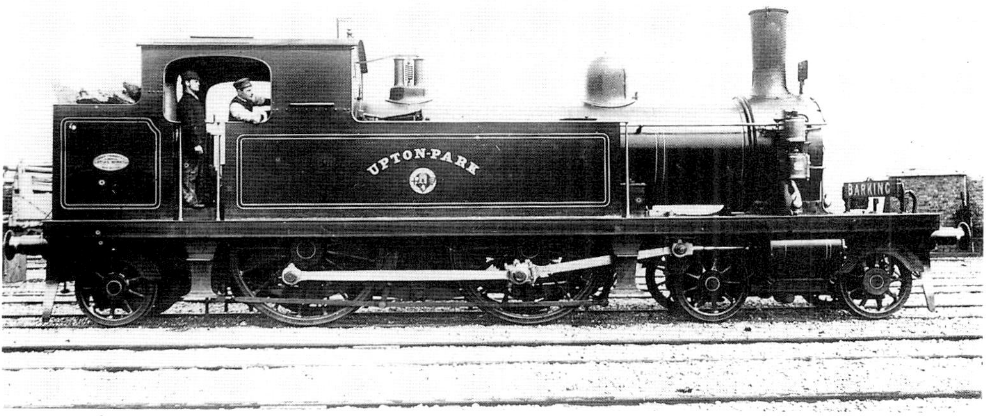

Plate 6: The off-side of the 'No. 1' class, No. 6 *Upton Park* built by Sharp, Stewart in 1880; the Westinghouse brake pump on the side of the smokebox was a feature of all LT&SR engines. *Real Photographs*

Plate 7: The later 'No. 1' class engines built by Sharp, Stewart had snap-head rivets showing the outlines of the side and bunker water tanks. No. 7 *Thames-Haven*, was built in 1881; the name was not normally hyphenated as here. *Real Photographs*

Plate 8: The 4-4-2T *St Pancras*, LT&SR No. 31, built in 1892 by Nasmyth Wilson; note the stove-pipe chimney carried by all the 'No. 1' class until replaced by cast chimneys, also the double whistle. The Gravesend headboard refers to a journey that finishes on a ferry.

F. Moore

Chapter Two

Locomotives

As stated earlier, until 1880 the Great Eastern Railway supplied locomotive power for working the LT&SR, but from that year, the company found all its own locomotives. The first to be delivered were the 'No. 1' class, which set the norm for most of the subsequent stock. These were 4-4-2 tank engines; who was actually responsible for the design is a moot point. They were built by Sharp, Stewart & Co. of Manchester, and in dimensions and details were very reminiscent of Adams' first 4-4-0 tanks (later rebuilt to 4-4-2) for the London & South Western Railway; it cannot be established that Adams was actually the producer of the design, nevertheless it seems fairly certain that he had a hand in the drawings somewhere. Thomas Whitelegg came from Sharp's works with the first batch of the 'No. 1' class locomotives, to oversee the delivery and preparations for service, and he remained at Plaistow Works as the LT&SR's first locomotive superintendent, which post he held until 1910, being succeeded by his son Robert. He (Robert) saw the take-over by the Midland Railway in 1912, but remained in charge at Plaistow until 1918, when he left to superintend the locomotive stock of the G&SWR in Scotland.

All classes except the last 4-4-2 and the 4-6-4 tanks were fitted with safety chains on the buffer beams; these were removed from some engines when they were rebuilt, and were generally removed about 1910.

No. 1 Class

The 'No. 1' class were simple and straightforward engines, and no matter who was responsible for their design, they proved eminently suited to the requirements of the LT&SR. The boiler, outside cylinders, and bogie were pure Adams in design, while the cab, tanks and bunker were of Sharp, Stewart's own particular pattern; the two separate designs married together very well. They had single slide bars and cottered big ends to the connecting rods; the coupling rods were of rectangular section, but were later altered to 'I' section as a result of experiencing one or two bent rods in service. Originally all the class had rather austere stove-pipe chimneys, subsequently replaced by slightly shorter cast chimneys with a rolled top, designed by Thomas Whitelegg. This type of chimney became standard, though the length varied with different classes.

The cylinders were 17 in. x 26 in., coupled wheels 6 ft 1 in., bogie and trailing wheels 3 ft 1 in. Wheelbase was 6 ft 6 in. + 7 ft 1 in. + 8 ft 6 in. + 7 ft 3 in., total 29 ft 4 in. The boiler was 4 ft 1 in. in diameter and 10 ft 6 in. long, containing 200 x 1⅝ in. tubes, giving a heating surface of 922 sq. ft, to which the firebox added 98 sq. ft, the total thus being 1,020 sq. ft. A round topped firebox, with outside length of 6 ft and grate area 17¼ sq. ft was fitted. Working pressure was 160 lb. and weight in working order 56 tons 2 cwt. The tanks held 1,260 gallons of water, and the bunker 2 tons of coal. When delivered, the engines had steam and hand brakes only, the Westinghouse air brake being added from 1885 onwards, when it was adopted as standard. There were 12 engines in the 1880 batch, and a further six, Nos. 13-18 were added in 1881, The class as a whole ultimately numbered 36 engines, Nos. 19-30 coming in 1884 from Sharp, Stewart, and finally Nos. 31-36 in 1892, this time from another Manchester builder, Nasmyth, Wilson & Co. For some reason, Nos. 19-30 were built with snap-head rivets in the tanks and bunkers, instead of the countersunk type, and looked rather less neat than the rest of the class. The Nasmyth engines had cylinder barrels ⅜ in. shorter between covers than the rest, which proved a bad feature in course of service; when the connecting rod brasses were adjusted, with the effect of slightly lengthening the rods, some damage to the cylinder covers was produced by the pistons striking them. In due course the six engines had new cylinders fitted with the standard length.

Two engines, Nos. 23 and 29, were modified as condensing engines in 1902 for working in the District Railway tunnels on the Whitechapel & Bow section. These had cast iron chimneys fitted, and in 1904 'wrap-over' cabs (*see figures 1 and 2*). Nos. 3 and 7 were later also altered in similar fashion; water pumps were fitted on the near side of the smokebox. Though the steam services in the tunnels lasted only three or four years, one of the condensing engines was always employed on the Plaistow Works shunt, so that it could be easily available in the case of a failure of the electricity supply on the District line. These four engines were always painted black after alteration instead of the standard green.

All the Tilbury engines, with the exception of the two tender engines and those delivered directly to the Midland Railway in 1912, were named. The names were carried in a curve over the coat of arms in the centre of the side tanks. Numbers were carried on the buffer beams only. The names began by using those of stations on the LT&SR; when these ran short, stations to which the company had running powers were used, and finally the names of parishes in that part of Essex through which the railway passed. Even so, the last named engine, No. 82, had to be named after a large rock in the Thames off Southend - *Crowstone*. Twelve engines ordered by the LT&SR, in 1912 were delivered in Midland livery just about the time of the takeover; these were not named, and there would have been much head scratching to unearth twelve new names! Even so, there were a few names which sounded faintly ridiculous on a railway locomotive - *Black Horse Road* for example -

I

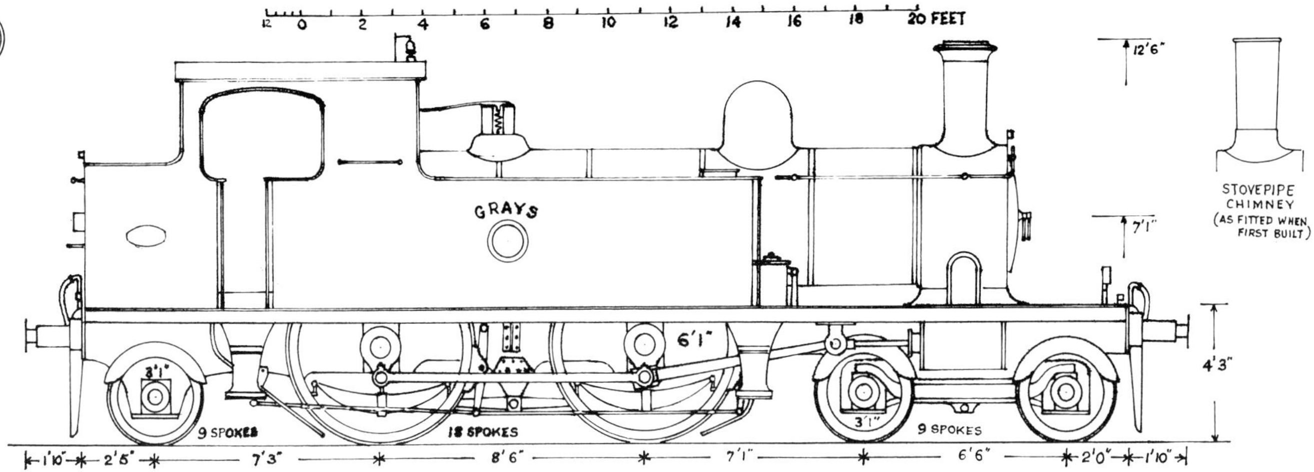

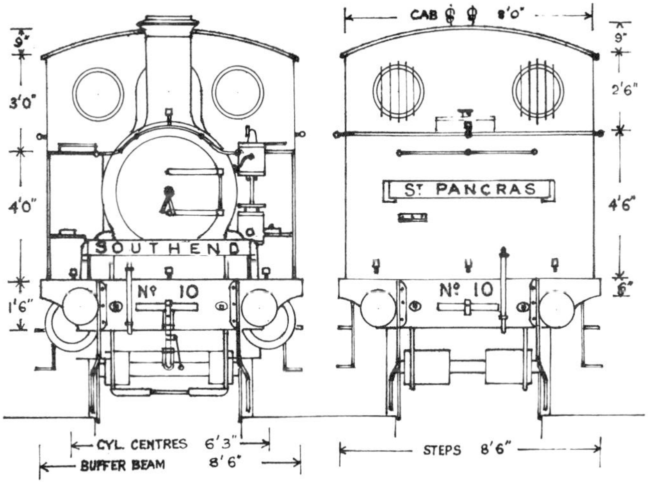

I CLASS

CYLS. 17" x 26"

36 BUILT 1880-1892

		BUILDERS	M.R.	L.M.S. 1923			BUILDERS	M.R.	L.M.S. 1923
1	SOUTHEND	SS 2880 1880	2110	2200	19	DAGENHAM	SS 3217 1884	2128	2193
2	GRAVESEND	" 2881 "	2111	2201	20	HORNCHURCH	" 3218 "	2129	2194
3	TILBURY	" 2882 "	2112	2202	21	UPMINSTER	" 3219 "	2130	2195
4	BROMLEY	" 2883 "	2113	2203	22	EAST HORNDON	" 3220 "	2131	2196
5	PLAISTOW	" 2884 "	2114	2204	23	LAINDON	" 3221 "	2132	2197
6	UPTON PARK	" 2885 "	2115	2205	24	OCKENDON	" 3222 "	2133	2198
7	BARKING	" 2886 "	2116	2206	25	STIFFORD	" 3223 "	2134	2199
8	RAINHAM	" 2887 "	2117	2207	26	WEST THURROCK	" 3224 "	2135	2056
9	PURFLEET	" 2888 "	2118	2208	27	WHITECHAPEL	" 3225 "	2136	2057
10	GRAYS	" 2889 "	2119	2209	28	ROMFORD	" 3226 "	2137	2058
11	STANFORD	" 2890 "	2120	2210	29	STEPNEY	" 3227 "	2138	2059
12	PITSEA	" 2891 "	2121	2211	30	FENCHURCH	" 3228 "	2139	2060
13	BENFLEET	" 2969 1881	2122	2212	31	ST PANCRAS	NW 425 1892	2140	2061
14	LEIGH	" 2970 "	2123	2213	32	LEYTON	" 426 "	2141	2062
15	EAST HAM	" 2971 "	2124	2214	33	WANSTEAD	" 427 "	2142	2063
16	LOW STREET	" 2972 "	2125	2190	34	TOTTENHAM	" 428 "	2143	2064
17	THAMES HAVEN	" 2973 "	2126	2191	35	WEST HAM	" 429 "	2144	2065
18	SHOEBURYNESS	" 2974 "	2127	2192	36	WALTHAMSTOW	" 430 "	2145	2066

SS = SHARP, STEWART & Co. NW = NASMYTH, WILSON & Co.

though they were actually named after stations. There was a little exchanging of names about 1909, for various reasons.

After the Midland took charge, various modifications were made as the engines went through shops - for all major overhauls they now went to Derby. All the class acquired Midland chimneys, and several a complete set of Midland boiler mountings. Some were fitted with extended smokeboxes and flat Deeley smokebox doors (which were terrors to keep tight) and various odd small fittings. The class were the first Tilbury engines to be withdrawn beginning in June 1930 with Nos. 9 and 10, and ending in October 1935 with No. 22, which was the last of the class in service. Most Tilbury engines were renumbered at least twice by the Midland Railway, first in 1912 with a Midland number, then when they came under LMS control, there were other re-numberings, usually to free their numbers for new stock. This treatment has led to a considerable amount of confusion.

					Re-numberings			
No.	*Name*	*Makers*	*Date*	*Rebuilt*	*1912*	*1923*	*1929*	*wdn.*
1	*Southend*	SS 2880	1880	1901	2110	2200	2077	11/30
2	*Gravesend*	SS 2881	1880	1903	2111	2201	2078	9/35
*3	*Tilbury*	SS 2882	1880	1902	2112	2202	2079	9/35
4	*Bromley*	SS 2883	1880	1903	2113	2203	2080	12/32
5	*Plaistow*	SS 2884	1880	1902	2114	2204	2081	8/30
6	*Upton Park*	SS 2885	1880	1902	2115	2205	2082	12/32
*7	*Barking*	SS 2886	1880	1898	2116	2206	2083	9/35
8	*Rainham*	SS 2887	1880	1902	2117	2207	2084	12/32
+9	*Purfleet*	SS 2888	1880	1903	2118	2208	2085	6/30
10	*Grays*	SS 2889	1880	1903	2119	2209	2086	6/30
11	*Stanford*	SS 2890	1880	1903	2120	2210	2087	11/30
12	*Pitsea*	SS 2891	1880	1902	2121	2211	2088	12/30
+13	*Benfleet*	SS 2969	1881	1906	2122	2212	2089	11/32
14	*Leigh*	SS 2970	1881	1903	2123	2213	2090	9/35
15	*East Ham*	SS 2971	1881	1904	2124	2214	2091	12/34
16	*Low Street*	SS 2972	1881	1904	2125	2190	2067	7/34
17	*Thames Haven*	SS 2973	1881	1903	2126	2191	2068	9/35
+18	*Shoeburyness*	SS 2974	1881	1901	2127	2192	2069	8/30
19	*Dagenham*	SS 3217	1884	1904	2128	2193	2070	9/35
20	*Hornchurch*	SS 3218	1884	1901	2129	2194	2071	11/32
21	*Upminster*	SS 3219	1884	1907	2130	2195	2072	9/35
+22	*East Horndon*	SS 3220	1884	1904	2131	2196	2073	10/35
*23	*Laindon*	SS 3221	1884	1902	2132	2197	2074	12/30
24	*Ockenden*	SS 3222	1884	1899	2133	2198	2075	9/35
25	*Stifford*	SS 3223	1884	1900	2134	2199	2076	9/35
26	*West Thurrock*	SS 3224	1884	1905	2135		2056	12/32
27	*Whitechapel*	SS 3225	1884	1901	2136		2057	12/32
28	*Romford*	SS 3226	1884	1903	2137		2058	9/35
*29	*Stepney*	SS 3227	1884	1902	2138		2059	9/32

②

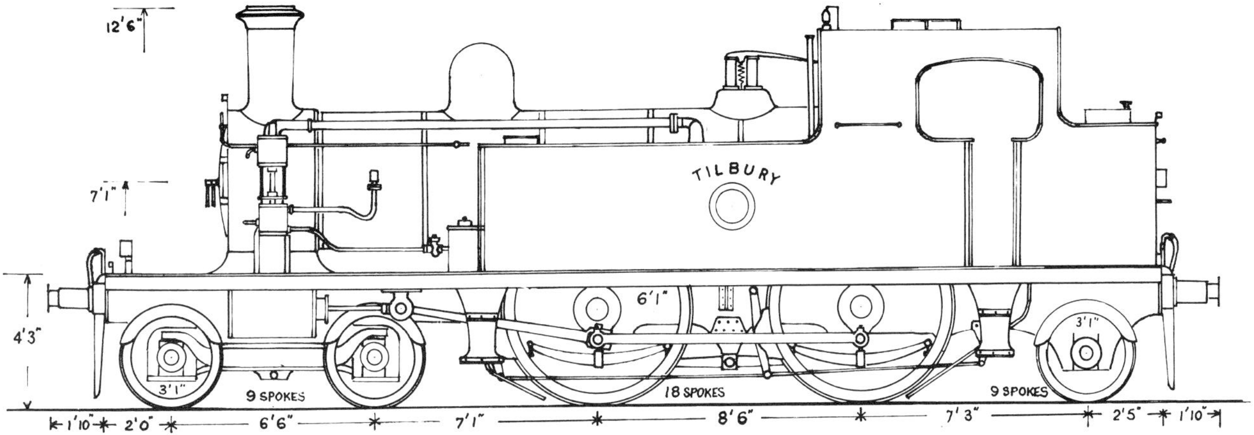

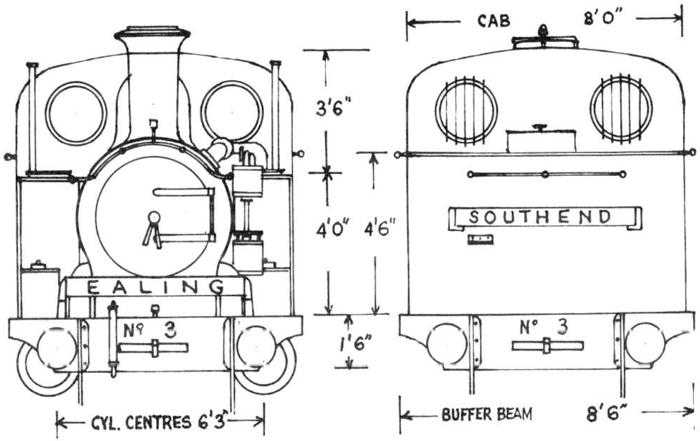

I CLASS

4 ENGINES MODIFIED WITH CONDENSING GEAR. 1902.

Nos. 3. 7. 23. 29.

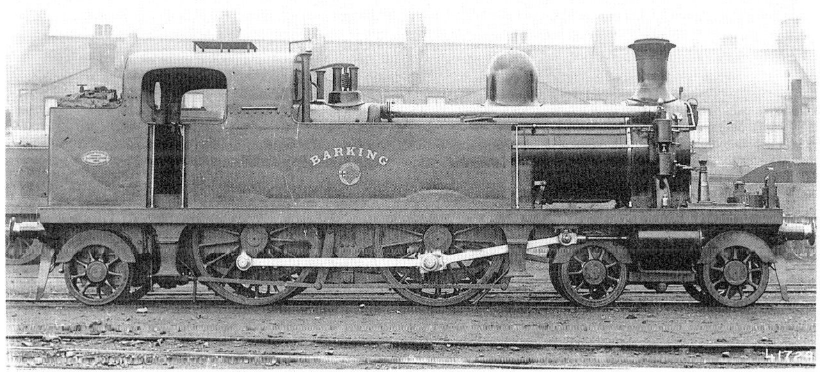

Plate 9: No. 7 *Barking*, after the fitting of condensing gear in 1902 to allow working through the District Railway tunnels; note the wrap-over cab with its ventilator.

Real Photographs

Plate 10: A front view of 'No. 1' class No. 5 *Plaistow* after fitting of a cast chimney; note the safety chains. *Real Photographs*

Re-numberings

No.	*Name*	*Makers*	*Date*	*Rebuilt*	*1912*	*1923*	*1929*	*wdn.*
30	*Fenchurch*	SS 3228	1884	1907	2139		2060	9/34
31	*St Pancras*	NW 425	1892	1904	2140		2061	12/32
32	*Leyton*	NW 426	1892	1905	2141		2062	12/32
33	*Wanstead*	NW 427	1892	1905	2142		2063	9/35
34	*Tottenham*	NW 428	1892	1905	2143		2064	12/32
35	*West Ham*	NW 429	1892	1903	2144		2065	12/32
36	*Walthamstow*	NW 430	1892	1904	2145		2066	8/32

SS = Sharp, Stewart & Co. NW = Nasmyth, Wilson & Co.

* Rebuilt with condensing gear and new cab, 1902

\+ Renamed 1911, to *Black Horse Road* (9); *Commercial Road* (13); *Burdett Road* (18); *Tilbury Docks* (22)

No. 37 Class

These were the first engines designed by Thomas Whitelegg, although it must be said that they were based largely on the 'No. 1' class. However, there were several increased dimensions in this class of twelve engines, and they were the only class to be extensively rebuilt subsequently. For all the 4-4-2 tanks which followed this class, the wheelbase, and length of boiler and firebox remained unaltered from those of the '37' class. The lengths of tanks and bunkers remained the same also, increased capacity being obtained by altering the height. The class were built in two batches of six in 1897 and 1898, Nos. 37-42 by Sharp, Stewart & Co., and 43-48 by Dübs & Co. of Glasgow. As originally built the engines had 18 in. x 26 in. cylinders, and 6 ft 6 in. coupled wheels, the bogie and trailing wheels being 3 ft 6 in. The wheelbase was increased to 7 ft 0 in. + 7 ft 0½ in. + 8 ft 9 in. + 8 ft 0 in., total 30 ft 9½ in. The boiler was 4 ft 2 in. in diameter and 10 ft 6 in. long, pitched at 7 ft 6 in., and contained 201 x 1⅝ in. tubes. Heating surface was 929 + 117, total 1,046 sq. ft; the firebox was 6 ft 9 in. long outside, with a grate area of 19.77 sq. ft. The working pressure was 170 lb., and weight in working order 63 tons 3 cwt. Water capacity was 1500 gallons, and coal 2¼ tons.

The whole class was rebuilt between 1905 and 1911, the amount of rebuilding being quite extensive. The boiler was 7 in. larger in diameter, at 4 ft 9 in., and in consequence was pitched at 8 ft 3 in.; the length and working pressure remaining the same. However, the heating surface was considerably higher, the number of tubes being increased to 212, giving 980 sq. ft, and with the 119 in the firebox, the total was 1099 sq. ft. Firebox length and grate area were unchanged. New cylinders were fitted, 19 in. in diameter and with 26 in. stroke. Owing to the higher pitch of the boiler, the tanks had to be moved outwards, and since they came out to the full width of the footplate, a kind of channel was formed at footplate level the length of the tank, so that the toe

Plate 11: A side view of 'No. 1' class 4-4-2T No. 34 *Tottenham* built by Nasmyth, Wilson & Co. in 1892 and withdrawn in 1932. *H.C. Casserley*

Plate 12: 4-4-2T No. 2076 (ex-LT&SR No. 25 *Stifford*) seen here at Tilbury on 22nd August, 1931. This locomotive was built by Sharp, Stewart & Co. and was withdrawn in 1935. One of the 19-30 batch, in which snap-head rivets were used in the tanks and bunker - the only batch to have this feature. *H.C. Casserley*

3

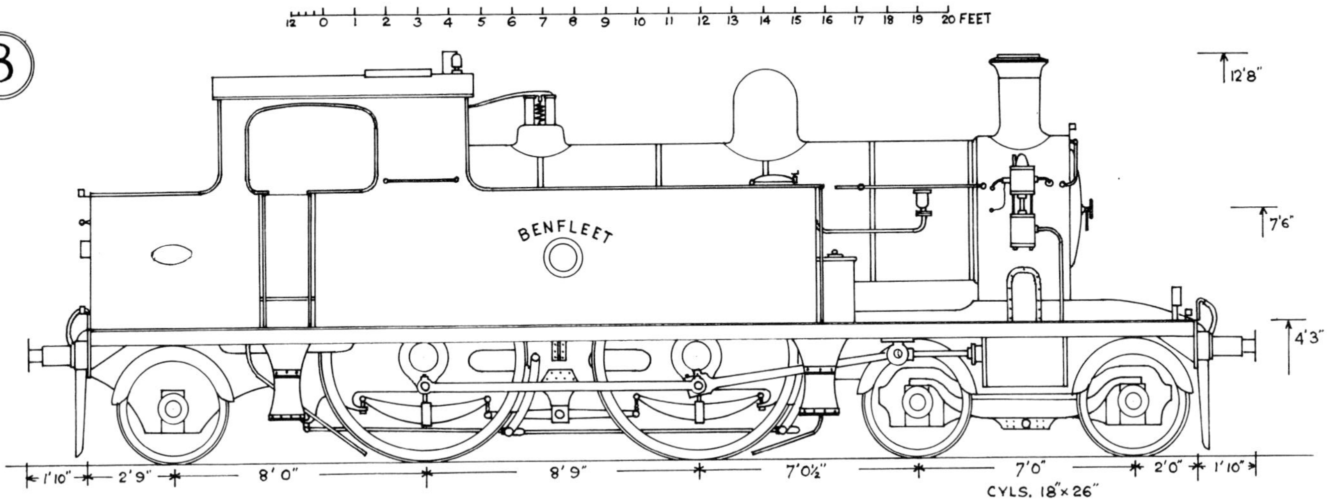

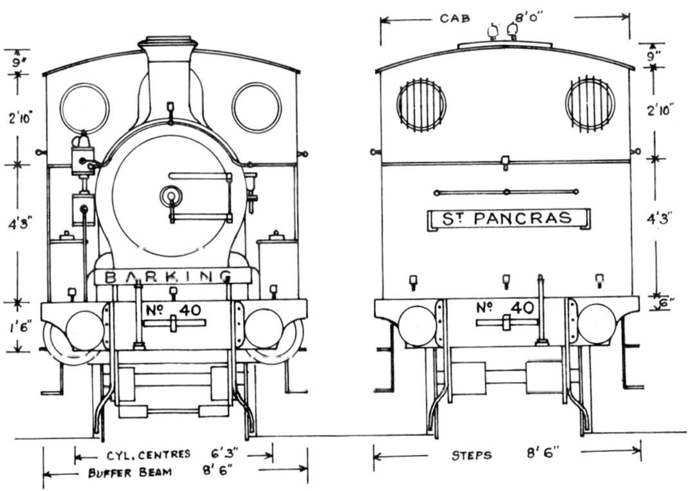

37 CLASS.

12 BUILT 1897-1898. Nos. 37-48.

	MAKERS	M.R.	L.M.S.		MAKERS	M.R.	L.M.S.
37 WOODGRANGE	SS 4245 1897	2146	2135	43 GREAT ILFORD	D 3666 1898	2152	2141
38 WESTCLIFF	SS 4246 "	2147	2136	44 PRITTLEWELL	D 3667 "	2153	2142
39 FOREST GATE	SS 4247 "	2148	2137	45 SHOEBURYNESS X	D 3668 "	2154	2143
40 BENFLEET X	SS 4248 "	2149	2138	46 SOUTHCHURCH	D 3669 "	2155	2144
41 LEYTONSTONE	SS 4249 "	2150	2139	47 STRATFORD	D 3670 "	2156	2145
42 EAST HORNDON X	SS 4250 "	2151	2140	48 LITTLE ILFORD	D 3671 "	2157	2146

SS = SHARP, STEWART & Co. D = DÜBS & Co

X = Renamed 1911. Original names were (40) BLACK HORSE ROAD. (42) COMMERCIAL ROAD. (45) BURDETT ROAD.

of a boot could be inserted, to enable the crew to pass along the engine when necessary. The coupled wheel splashers inside the tanks also had to be moved inwards. New 'wrap-over' cabs were also fitted, though this was not much of an improvement, since the fact of the cab cut-out projecting into the curve of the roof allowed rainwater to drip on to the top of the tank, forming a convenient puddle for the unwary engineman to put his elbow into. Steam reversing gear was also fitted, though this was viewed with mixed feelings by the drivers; some welcomed it, but others disliked it. It was a bit temperamental, and could be awkward at times, so perhaps the mixed reception from the drivers was understandable.

New boiler mountings were fitted, of reduced height, owing to the larger boiler, both the dome and safety valve covers being beaten up by hand, so that no two of them were exactly alike. Another innovation was the fitting of doors to the tops of the bunkers, these being fixed to the top of an additional raised portion with curved sides. The idea of this was to prevent the rather dusty coal used from being blown all over the place, when working bunker first, also to discourage over filling of the bunker. This feature appeared on all subsequent engines built, and gradually extended to the older classes. As rebuilt, the 'No. 37' class had a water capacity of 1,728 gallons, and the weight in working order was increased to 65 tons 7 cwt.

No. 37 was the first to be rebuilt, in 1905, followed by Nos. 39 and 43 in 1907, 47 in 1908, and 38 and 45 in 1909. Three were dealt with in 1910, Nos. 41, 42, and 46, and the last three, Nos. 40, 44 and 48 in 1911. Two Nos. 38 and 40, were rebuilt a second time in 1920. Those rebuilt prior to 1910 were given an experimental livery of lavender blue, with black borders, and lined in white. However this did not stand up to weathering at all well, and the engines were gradually repainted in the standard light green. The rebuilt class '37' engines were put into the top link, several of them being shedded at Shoeburyness, and worked the heaviest expresses along with the '79' class which came out later and were almost identical. In Midland days they received all sorts of modifications such as Derby boiler mountings, smokebox doors, and extended smokeboxes, in various combinations. One, No. 44, acquired a complete Midland boiler with Belpaire firebox. The entire class was withdrawn during 1951.

					Re-numberings			
No.	*Name*	*Makers*	*Date*	*Rebuilt*	*1912*	*1929*	*1946*	*wdn.*
37	*Woodgrange*	SS 4245	1897	1905 1918	2146	2135	1953	7/51
38	*Westcliff*	SS 4246	1897	1909 1920	2147	2136	1954	7/51
39	*Forest Gate*	SS 4247	1897	1907	2148	2137	1955	3/51
+40	*Black Horse Road*	SS 4248	1897	1911 1920	2149	2138	1956	5/51

Plate 13: No. 47 *Stratford* was one of the second batch of class '37' engines, built by Dübs in 1898. *Real Photographs*

Plate 14: This '37' class engine *Great Ilford* was rebuilt in 1909 with larger boiler and 19 in. X 26 in. cylinders, also wrap-over cab. The whole class was similarly treated between 1905 and 1911. *Real Photographs*

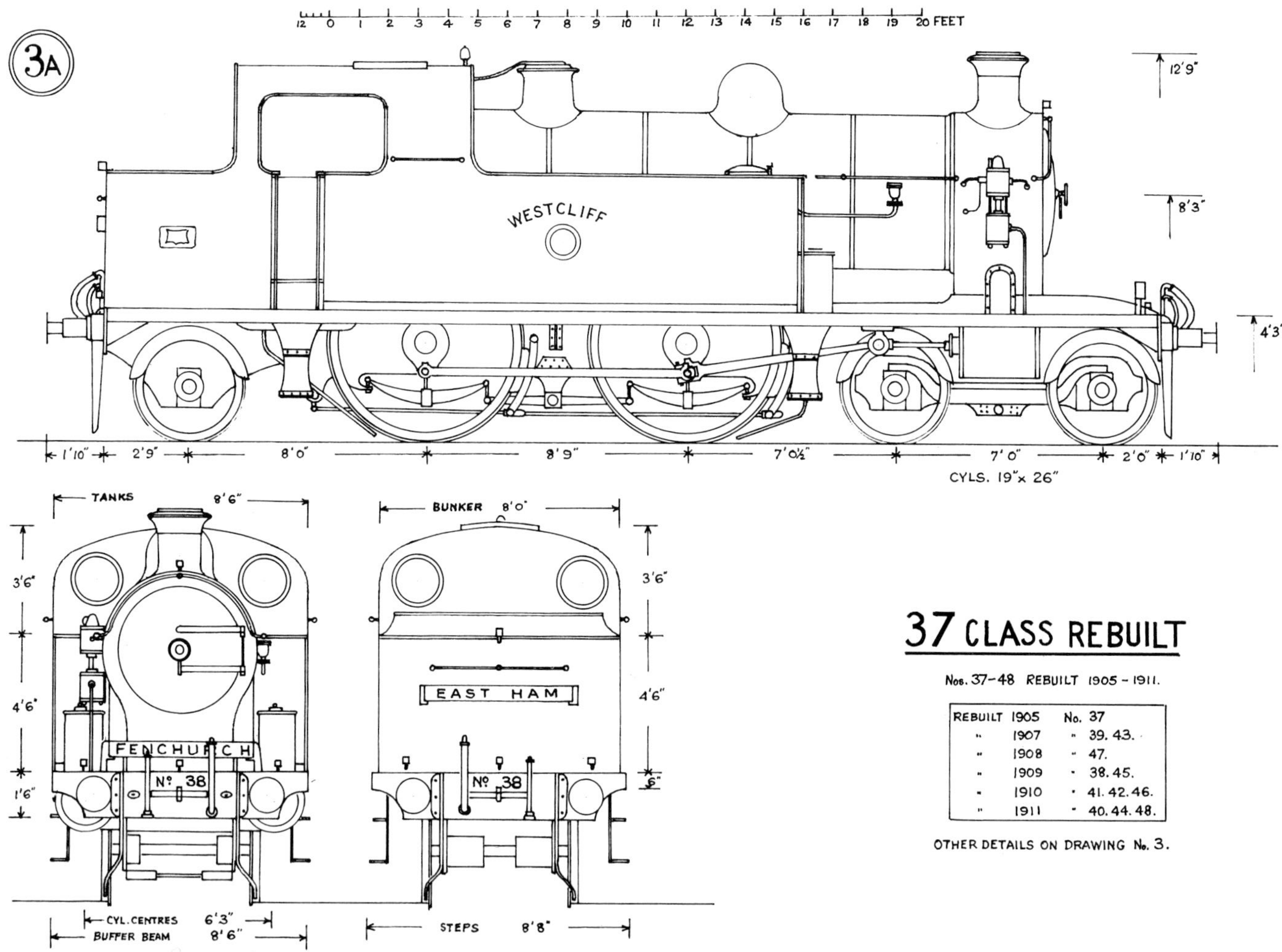

37 CLASS REBUILT

Nos. 37–48 REBUILT 1905 – 1911.

REBUILT	1905	No. 37
"	1907	" 39. 43.
"	1908	" 47.
"	1909	" 38. 45.
"	1910	" 41. 42. 46.
"	1911	" 40. 44. 48.

OTHER DETAILS ON DRAWING No. 3.

Re-numberings

No.	Name	Makers	Date	Rebuilt	1912	1929	1946	wdn.
41	*Leytonstone*	SS 4249	1897	1910	2150	2139	1957	3/51
+42	*Commercial Road*	SS 4250	1897	1910	2151	2140	1958	12/51
43	*Great Ilford*	D 3666	1898	1907	2152	2141	1959	4/51
44	*Prittlewell*	D 3667	1898	1911 1914	2153	2142	1960	6/51
+45	*Burdett Road*	D 3668	1898	1909	2154	2143	1961	12/51
46	*Southchurch*	D 3669	1898	1910	2155	2144	1962	2/51
47	*Stratford*	D 3670	1898	1908	2156	2145	1963	2/51
48	*Little Ilford*	D 3671	1898	1911	2157	2146	1964	3/51

SS = Sharp, Stewart & Co. Glasgow D = Dübs & Co. Glasgow

\+ Renamed 1911, to *Benfleet* (40); *East Horndon* (42); *Shoeburyness* (45)

No. 49 Class

These were a pair of oddments - Nos. 49 and 50 - the only tender engines the company ever had. Since tank engines were perfectly capable of handling anything the system required, the acquisition of these two was apparently an aberration, though the fact that they were obtained at a bargain price may have had something to do with it. They had been built by Sharp, Stewart & Co. in 1899 (Works Nos. 4420/1) as part of an order for the Ottoman Railways in Turkey, but for some reason were left on the builders' hands. As delivered to Plaistow the cabs had a very wide unglazed opening on each side; before going into service this was filled in and provided with a normal carriage window in the centre. None of the dimensions of these two 0-6-0s fitted in with those of the 4-4-2 tanks, and they remained totally non-standard for the whole of their career. At one period - between 1910 and 1916 - they seem to have struck a bad patch, all sorts of odd things going wrong, but apparently they recovered from this, and remained in service until well into the LMS period, No. 49 being withdrawn in December 1933, as No. 2898, and No. 50, then renumbered 22899, lasted until February 1936. Both had acquired Midland pattern Belpaire boilers in the mid-1920s.

As built, their dimensions were as follows; Coupled wheels 4 ft 6½ in., inside cylinders 18 in. x 24 in., wheelbase 7 ft 5 in. + 7 ft 11 in., total 15 ft 4 in. The boiler was 4 ft 4 in. in diameter and 10 ft 0 in. long, pitched at 7 ft 0 in., and contained 216 x 1¾ in. tubes, which gave a heating surface of 1,018 sq. ft. With the 96 sq. ft provided by the firebox, the total heating surface was 1,114 sq. ft. The firebox was 4 ft 10 in. long, and had a grate area of 16 sq. ft. Working pressure was 150 lb. The tender ran on six 3 ft 6½ in. wheels, on an equally-divided base of 12 ft 0 in., and carried 2,640 gallons of water and 4 tons of coal. Neither of these engines was ever named. No. 49 spent a good deal of her career as the Engineering Department engine, based at Tilbury, and pottered around with ballast and sleeper wagons, while No. 50 was stationed at Plaistow and took her turn in the goods link with the tank engines. A peculiar feature of both engines was the Westinghouse main air reservoir, which appeared as an oblong tank mounted on the left hand footplate immediately in front of the cab. About 1914 they had their cylinders lined up to 16½ in. diameter. They were not particularly good engines, while pulling hard the crews likened their motion to bucking bronchos, but nevertheless they were retained in stock as long as suitable work could be found for them, even the Midland recognising that they must have had some good points, since they survived well into the 1930s.

No. 51 Class

There were 18 engines in this class, built in two batches, Twelve by Sharp, Stewart & Co. in 1900, and six by the North British Locomotive Co. in 1903, carrying numbers 51-68. They were based on the '37' class, but with a larger boiler, 4 ft 6 in. in diameter and 10 ft 6 in. long pitched at 7 ft 9 in. Cylinders were 18 in. x 26 in., coupled wheels 6 ft 6 in., and carrying wheels 3 ft 6 in. Wheelbase, firebox and most other mechanical details were standard with the '37' class. The boiler contained 201 x 1⅝ in. tubes, with a heating surface of 929 sq. ft, which with the firebox contribution of 117 sq. ft gave a total of 1,046 sq. ft. The grate area was 19.77 sq. ft. and working pressure 170 lb. The cabs were of the standard pattern with separate arc roof, but the tanks were slightly higher, holding 1900 gallons, while the bunker capacity was 2½ tons. All the class were dual fitted, the vacuum brakes for working Midland stock to and from St Pancras, and No. 61 Kentish Town spent practically the whole of her career until 1923 stabled in the Midland running shed bearing the same name.

No. 64 (as MR No. 2171) came back from a general overhaul at Derby in 1921 with ½ in. sideplay in all her axleboxes instead of the usual ⅛ in. When put back into service, she felt most unsafe, floating about sideways in a most erratic manner, and was reported by every driver who took her out. As a result, she was relegated to stopping trains, until Derby condescended to take her back again, - which for LTSR engines was a minimum of twelve months. On one occasion, due to an oversight she was put on an express, and taking Pitsea curve at well over 50 mph, the chief permanent way inspector, who was riding in the front van, shot headfirst across the van floor. He reported the occurrence, and it blew up into a full scale departmental row, with the Locomotive Department trying to cover up for the engine's vagaries. However, the upshot of this was that No. 2171 went back to Derby as a 'special case' and had her axleboxes restored to normal.

④

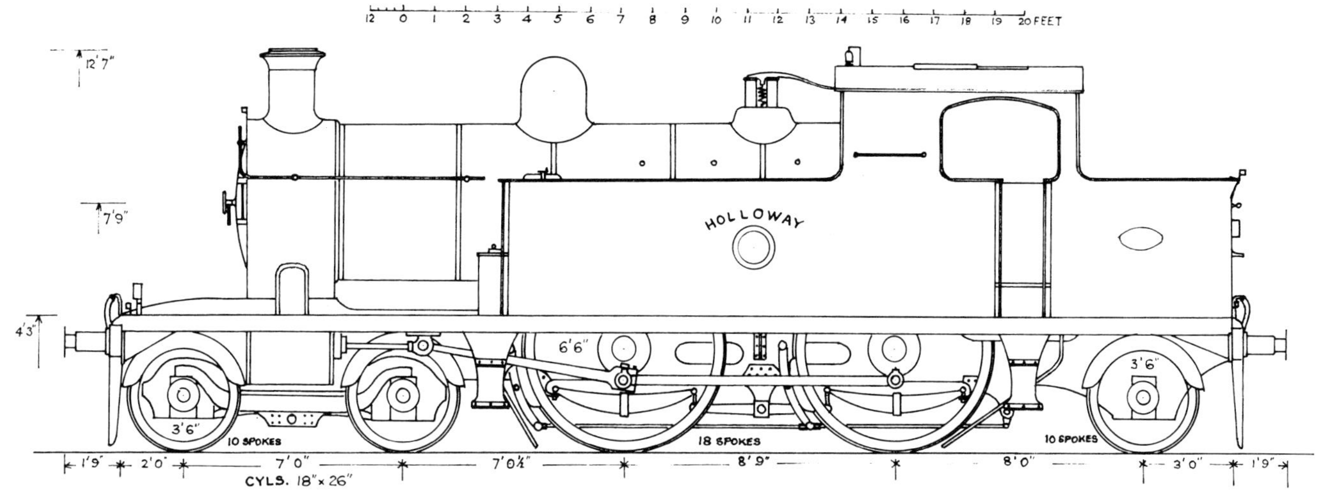

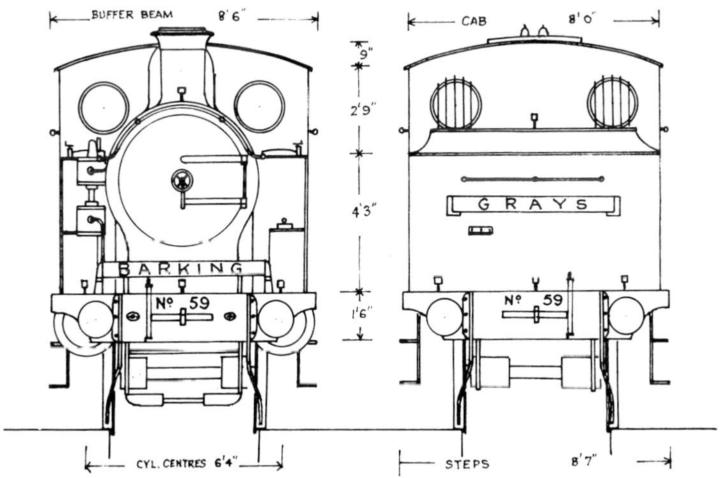

51 CLASS.

18 BUILT 1900-1903 Nos. 51-68.

	Builders.	M.R.	L.M.S.		Builders	M.R.	L.M.S.
51 Purfleet ×	SS 4653 1900	2158	2092	60 Highgate ×	SS 4662 1900	2167	2101
52 Wennington	SS 4654 〃	2159	2093	61 Kentish Town	SS 4663 〃	2168	2102
53 Stepney Green	SS 4655 〃	2160	2094	62 Camden ×	SS 4664 〃	2169	2103
54 Mile End	SS 4656 〃	2161	2095	63 Mansion House	NBL 15741 1903	2170	2104
55 Bow Road ×	SS 4657 〃	2162	2096	64 Charing Cross	NBL 15745 〃	2171	2105
56 Harringay	SS 4658 〃	2163	2097	65 Victoria	NBL 15746 〃	2172	2106
57 Crouch Hill	SS 4659 〃	2164	2098	66 Earls Court	NBL 15747 〃	2173	2107
58 Hornsey ×	SS 4660 〃	2165	2099	67 Westminster	NBL 15748 〃	2174	2108
59 Holloway	SS 4661 〃	2166	2100	68 Mark Lane	NBL 15749 〃	2175	2109

BUILDERS. SS = SHARP STEWART & Co. NBL = NORTH BRITISH LOCOMOTIVE Co.

X. THESE ENGINES WERE RENAMED. No. 55 IN 1903, THE OTHERS IN 1911.

ORIGINAL NAMES WERE — 51 TILBURY DOCKS.
55 WELLINGTON ROAD.
58 HORNSEY ROAD.
60 HIGHGATE ROAD.
62 CAMDEN ROAD.

Plate 15: The '51' class was almost identical with the '37' class; *Victoria* No. 65 was built by North British in 1903. *Real Photographs*

Plate 16: '51' class 4-4-2T No. 2101 (ex-LT&SR No. 60 *Highgate Road*) standing next to sister engine No. 2103 (ex-LT&SR No. 62 *Camden Road*). *H.C. Casserley*

The '51' class were a very free-running lot, and though they were just not capable of dealing with the heaviest trains (which were usually hauled by the rebuilt '37' class) they were often preferred for the normal expresses, and could always maintain time. As to their later days, the Midland adorned them with the usual Derby features, and in addition replaced the original wooden cab roofs with steel plate, resulting in a cosy and comfortable cab being transformed into a dank dripping mass of condensation. No. 2170 (former No. 63) ran for a considerable time in a hybrid livery of large Midland numbers on the original Tilbury green background. Most of the class received extended smokeboxes with flat smokebox doors. A number were transferred away from the Tilbury section in LMS days to sheds in the Midlands, where they were not known, with the consequence that they were put on unsuitable jobs, and they gained a bad reputation with the crews through no fault of their own.

No.	*Name*	*Makers*	*Date*	*Rebuilt*	*1912*	*1929*	*1945*	*wdn.*
+51	*Tilbury Docks*	SS 4653	1900	1914	2158	2092	1910	9/48
52	*Wennington*	SS 4654	1900		2159	2093	1911	7/51
53	*Stepney Green*	SS 4655	1900	1921	2160	2094	1912	8/49
54	*Mile End*	SS 4656	1900		2161	2095	1913	8/49
+55	*Wellington Rd*	SS 4657	1900		2162	2096	1914	6/50
56	*Harringay*	SS 4658	1900		2163	2097	1915	3/51
57	*Crouch End*	SS 4659	1900		2164	2098	1916	3/51
+58	*Hornsey Road*	SS 4660	1900		2165	2099	1917	3/51
59	*Holloway*	SS 4661	1900	1923	2166	2100	1918	12/49
+60	*Highgate Road*	SS 4662	1900		2167	2101	1919	3/51
61	*Kentish Town*	SS 4663	1900		2168	2102	1920	8/49
+62	*Camden Road*	SS 4664	1900		2169	2103	1921	3/51
63	*Mansion House*	NB 15744	1903	1918	2170	2104	1922	8/51
·64	*Charing Cross*	NB 15745	1903		2171	2105	--	11/47
65	*Victoria*	NB 15746	1903		2172	2106	1923	6/49
66	*Earls Court*	NB 15747	1903		2173	2107	1924	12/49
67	*Westminster*	NB 15748	1903		2174	2108	1925	8/51
68	*Mark Lane*	NB 15749	1903		2175	2109	1926	3/51

SS = Sharp, Stewart & Co. Glasgow
NB = North British Locomotive Co., Glasgow

+ Renamed (51) *Purfleet* in 1911 (55) *Bow Road* in 1903 (58) *Hornsey* in 1911 (60) *Highgate* in 1911 (62) *Camden* in 1911

No. 69 Class

Up to 1903, all goods work had been undertaken by the 4-4-2 tanks, even though their 6 ft 6 in. coupled wheels were unsuitable for this kind of service.

There had only been the two 0-6-0 tender locomotives of the '49' class which were true goods engines, and these could not possibly cope with all the work. So Thomas Whitelegg designed a class of tank engines which would be more suitable for goods service, though at the same time he included as much material as possible which would be standard with the passenger engines. They had the same cylinders, boiler, trailing radial wheels, cab, and tanks (as far as length goes; they were slightly higher). The differences were in the coupled wheels, which were 5 ft 3 in. in diameter, and the wheelbase, 7 ft 7 in. + 9 ft 3 in. + 8 ft 0 in., total 24 ft 0 in. Water capacity was 1,914 gallons, and coal 2½ tons; weight in working order 64 tons 13¾ cwt.

The first six of these 0-6-2 tanks, Nos. 69-74, were built by the North British Locomotive Co. in 1903, Works Nos. 15750-5, and the second batch of four came from the same makers in 1906, with the Works Nos. 18504-7, and running Nos. 75-78. It may be said here that the North British Locomotive Co. was formed in 1903 by the amalgamation of three old-established firms, Sharp, Stewart & Co., Neilson, Reid & Co., and Dübs & Co., all with workshops in Glasgow. So by ordering from NBL the LT&SR were perpetuating its long association with Sharp, Stewart, who although they originated in Manchester, had moved to Glasgow in 1888. However, the association was not to continue, for the final batches of engines ordered by the LT&SR before the takeover came from Robert Stephenson & Sons, and Beyer, Peacock & Co.

Four more of the 0-6-2 tanks were on order when the Midland Railway took over, and in consequence were delivered in Midland goods black, wlth their new numbers 2190-2193, though they were allotted 83-86 in the Tilbury list. These were the Beyer, Peacock engines mentioned above. They did not have names, though the previous 10 of the class were named. These last four differed in minor details; they had steam sanding, which proved unreliable and was removed after about two years. They also had Holden & Brooke injectors in place of the usual Gresham & Craven; these were rather tricky to work compared with the Gresham & Craven type, and were therefore unpopular with the crews, though quite satisfactory in use. The four 1908 engines were fitted with Thomas Whitelegg's variable blast pipe, and the favourable results from this experiment led to all the tank engines except the No. '1' class being fitted with them. They were promptly removed by the LMSR.

The '69' class were good engines, though a trifle sluggish due to the cramped position of the slide valves between the cylinders, but they steamed well, and were capable of any amount of hard work. Though their boilers were the same as the passenger engines, they were not interchangeable owing to one or two minor differences; the clack boxes were on the front ring of the boiler, whereas the passenger engines had them in the cab. The steampipe for the Westinghouse pump came from the boiler just in front of

⑤

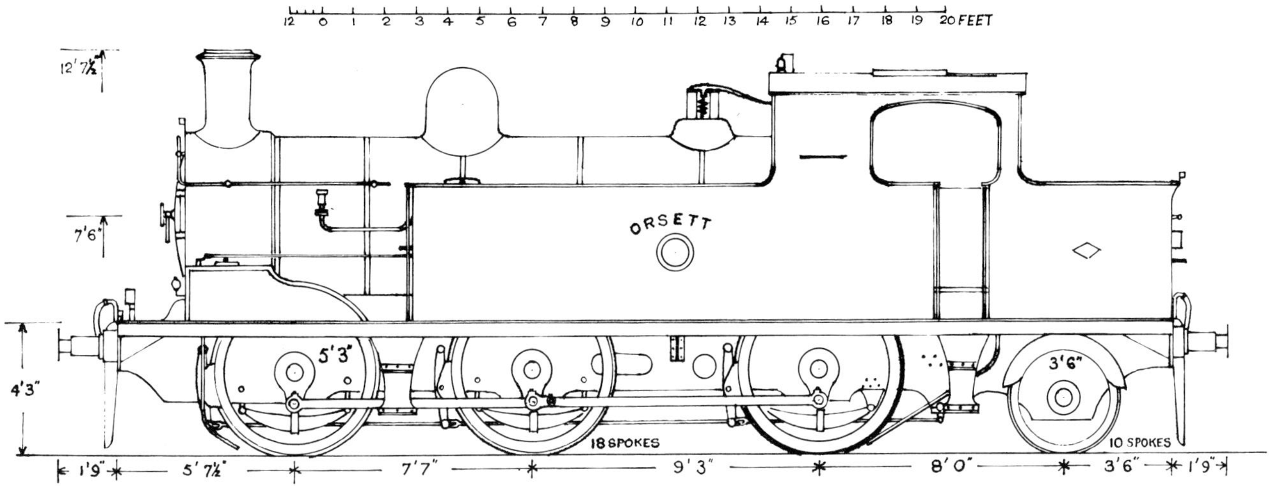

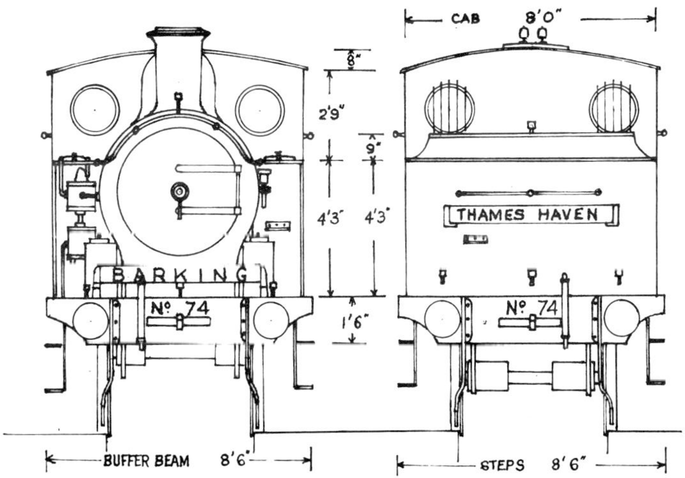

69 CLASS

14 BUILT 1903-1912 CYLS. 18" × 26"

	MAKERS	M.R.	LMS 1923		MAKERS	M.R.	LMS 1923
69 CORRINGHAM	NBL 15750 1903	2180	2220	76 DUNTON	NBL 18505 1905	2187	2227
70 BASILDON	" 15751 "	2181	2221	77 FOBBING	" 18506 "	2188	2228
71 WAKERING	" 15752 "	2182	2222	78 DAGENHAM DOCK	" 18507 "	2189	2229
72 HADLEIGH	" 15753 "	2183	2223	83 —	BP 5604 1912	2190	2230
73 CRANHAM	" 15754 "	2184	2224	84 —	" 5605 "	2191	2231
74 ORSETT	" 15755 "	2185	2225	85 —	" 5606 "	2192	2232
75 CANVEY ISLAND	" 18504 1906	2186	2226	86	" 5607 "	2193	2233

MAKERS — NBL = NORTH BRITISH LOCOMOTIVE Co.
BP = BEYER, PEACOCK & Co.

Plate 17: No. 69 *Corringham* was one of six '69' class goods tanks built by North British in 1903. Note that double whistles were still being fitted. The Gravesend board the engine is carrying would only have been used if it were pressed into passenger service.
Real Photographs

Plate 18: '69' class 0-6-2T No. 73 *Cranham*. Built by North British Locomotive Co., Glasgow in 1903.
Real Photographs

Plate 19: Sister engine to No. 73 was No. 76 *Dunton*, built later in 1908 and rebuilt in 1924. Finally scrapped in 1959. One can imagine the confusion that could arise with the travelling public, with a different name on the destination board at the front of the locomotive from that shown on the side tanks, as illustrated in this view.

H.C. Casserley

Plate 20: A further view of a '69' class locomotive, No. 78 *Dagenham Dock*, possibly photographed when new in 1908.

Real Photographs

the dome instead of from the smokebox, and (on the '69' class only) the omission of the steam cock for the vacuum ejector, since the goods tanks were not dual fitted.

The 1912 0-6-2 tanks came out with bunker extensions fitted with lids; these were also applied in a few cases to earlier classes, the object being to reduce flying dust, especially when working bunker first. The Midland later replaced these by coal rails.

When the Tilbury contract for engine oil expired, the goods tanks suddenly developed an alarming crop of hot big ends. The drivers all agreed this was due to the Midland brand of engine oil, which one driver described as a mixture of paraffin and fish glue. Also at that time there was a campaign on to 'use less oil'. After a few months the trouble disappeared as quickly as it had started, and was not heard of again.

After the LMSR took over in 1923 the 14 0-6-2 tanks were renumbered 2220-2233, but in 1939 were again renumbered back to 2180-93, to clear the 2200 series for new 2-6-4 tanks. They were finally renumbered again in 1949 to 1980-1993 for the same reason.

					Re-numberings				
No.	*Name*	*Makers*	*Date*	*Rebuilt*	*1912*	*1923*	*1939*	*1946*	*wdn.*
69	*Corringham*	NB 15750	1903		2180	2220	2180	1980	5/58
70	*Basildon*	NB 15751	1903		2181	2221	2181	1981	5/58
71	*Wakering*	NB 15752	1903	1922	2182	2222	2182	1982	2/59
72	*Hadleigh*	NB 15753	1903		2183	2223	2183	1983	2/59
73	*Cranham*	NB 15754	1903		2184	2224	2184	1984	2/59
74	*Orsett*	NB 15755	1903	1922	2185	2225	2185	1985	2/59
75	*Canvey Island*	NB 18504	1908		2186	2226	2186	1986	2/59
76	*Dunton*	NB 18505	1908	1924	2187	2227	2187	1987	3/59
77	*Fobbing*	NB 18506	1908		2188	2228	2188	1988	4/58
78	*Dagenham Dock*	NB 18507	1908		2189	2229	2189	1989	4/58
(83)		BP 5604	1912		2190	2230	2190	1990	3/59
(84)		BP 5605	1912		2191	2231	2191	1991	2/59
(85)		BP 5606	1912		2192	2232	2192	1992	2/59
(86)		BP 5607	1912		2193	2233	2193	1993	2/59

NB = North British Locomotive Co., Glasgow
BP = Beyer, Peacock & Co., Manchester

No. 79 Class

There were only four of this class, numbered 79-82, built by Robert Stephenson & Sons of Newcastle-on-Tyne in 1909, and were the last of the 4-4-2 type built for the LT&SR. For all practical purposes they were identical with the '37' class as rebuilt, with one or two minor alterations the chief of

⑥

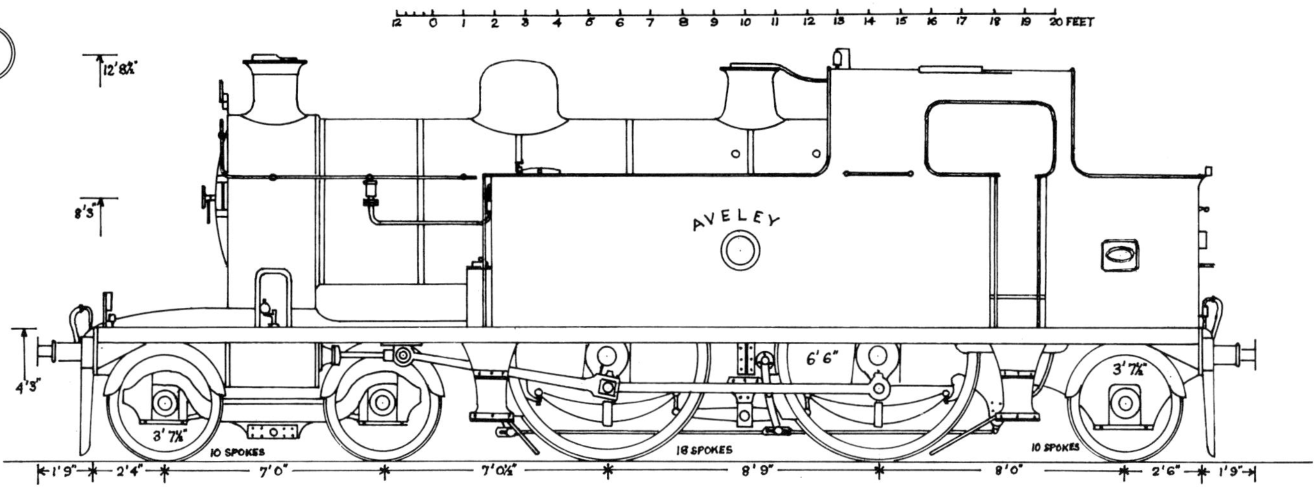

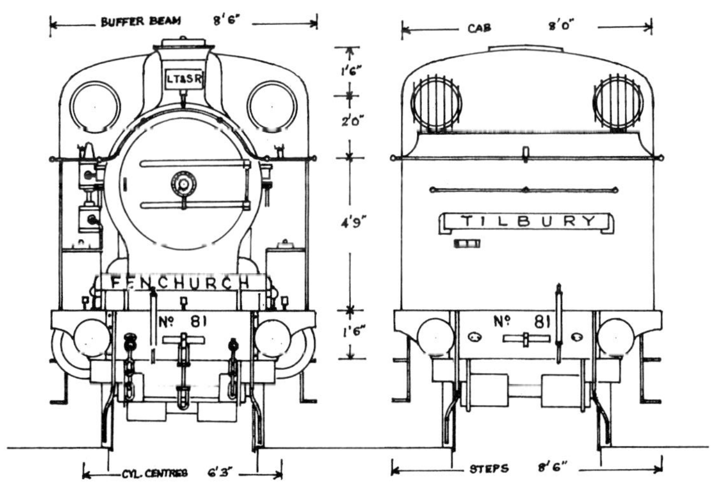

79 CLASS.

4 BUILT 1909.

	BUILDERS	M. R.	L. M. S.
79 RIPPLESIDE	RS 3366 1909	2176	2147
80 THUNDERSLEY	RS 3367 "	2177	2148
81 AVELEY	RS 3368 "	2178	2149
82 CROWSTONE	RS 3369 "	2179	2150

BUILDERS — RS = ROBERT STEPHENSON & SONS.

No. 80 WAS NAMED SOUTHEND-ON-SEA UNTIL 1910

which was further enlarged side tanks. They also had the wrap-over cabs as fitted to the '37' class. It is doubtful whether Stephensons made much profit from building these four engines, for three out of the four broke piston rods within four months of going into service, damaging the cylinders, and in the end Stephensons replaced all the eight cylinders and piston rods free of charge.

No. 80, *Thundersley*, was specially renamed *Southend-on-Sea* and exhibited at the White City International Exhibition later in 1909, where she gained a gold medal. In 1911, the same engine (with her original name restored) was specially decorated for the coronation of King George V; £600 was earmarked for this purpose. White busts of the new King and Queen were fitted on the front corners of the footplate, and a bust of Queen Alexandra was mounted on the rear of the bunker. The chimney was fitted with a nickel plated cap, and the safety valve cover was also nickel plated, as were the lagging bands of the cylinders and Westinghouse pump, while the boiler bands were of polished brass, sweated onto wider bands of polished steel, showing half an inch on either side. On each side of the boiler bands was a crimson lake edging, and a fine red line. The smokebox was fitted with polished steel bands; the hinge straps and handwheel were also polished, and the buffer faces were nickel plated. The footplate was edged with gilt chains, and a small ornamental fountain was fixed to the front of the footplate between the frames, with a water supply from the side tanks. After the engine returned from exhibition, it went into service still decorated, but only for about a month; most of the special decoration was then removed, though the boiler bands, chimney cap, safety valve cover and other polished bands remained until the engine's next visit to Derby, when it reverted to normal.

The cylinders of the '79' class were 19 in. x 26 in., coupled wheels 6 ft 6 in., and bogie and trailing wheels 3 ft 6 in.; the wheelbase was the same as the '37' class. The boiler was 4 ft 8 in. diameter and 10 ft 6 in. long, working at 170 lb., and had 217 x 1⅝ in. tubes. Heating surface was 1,003 + 119, total 1,122 sq. ft, grate area 19.77 sq. ft, and the firebox 6 ft 9 in. long. Water capacity was 1,926 gallons, and coal 2½ tons. Weight in working order was 69 tons 7½ cwt.

The '79' class became numbers 2176-2179 in the Midland list, which numbers they retained in the 1923 LMS list, but in 1929 they were altered to 2147-2150, and in 1949 they were again altered to 1965-1969. During 1923 to 1930 a further 35 were built, mostly at Derby, with Midland boilers and fittings. These were Nos. 2110-2119 (1923); 2120-2124 (1925); 2125-2134 (1927) and 2151-2160 (1930). (Note - the five built in 1925 were constructed by Nasmyth, Wilson & Co. (Works Nos. 1448-1452), the rest at Derby.)

Plate 21: *Thundersley* photographed when new in 1909 showing the livery and lining details clearly. The engine came into the '79' class as No. 80 and was built in 1909 by Robert Stephenson & Sons of Newcastle upon Tyne. It was not withdrawn until 1956.
Real Photographs

Plate 22: The '79' class was the last 4-4-2T series built before fusion with the Midland; No. 81 *Aveley* was the third of four built by Robert Stephenson in 1909.
Real Photographs

				Re-numberings			
No.	*Name*	*Makers*	*Date*	*1912*	*1929*	*1949*	*wdn.*
79	*Rippleside*	RS 3366	1909	2176	2147	1965	12/51
80	*Thundersley*	RS 3367	1909	2177	2148	1966	3/56
81	*Aveley*	RS 3368	1909	2178	2149	1967	12/52
82	*Crowstone*	RS 3369	1909	2179	2150	1968	3/51

RS = Robert Stephenson & Sons, Newcastle-on-Tyne,

No. 87 Class

Robert Whitelegg's only design for the LT&SR, was an entirely new departure from the traditional 4-4-2, but at the same time was a logical progression from it. The original design for this large 4-6-4 tank engine was considerably different from the engines actually built, the driving wheels being 5 ft 9 in. instead of the 6 ft 3 in. actually adopted. As originally drafted at Plaistow, they had boiler mountings similar to those of the '79' class, and a wrap-over cab slightly longer than on the 4-4-2s. As built, the boiler was larger in diameter than originally planned, necessitating shorter boiler mountings; the smokebox was longer and had a very deeply dished door, while the cab was entirely different in design, having straight side sheets, a smaller cut-out, and a high arc roof. A side window with curved top was fitted in the forward part of the cab side sheet. The bunker was also slightly shortened from the original design and had an extension with doors on top. As built, the outside cylinders were fitted with tail rods, which did not appear on the original drawings. Schmidt superheaters were fitted, though about 1920 one engine (No. 2100) had the apparatus removed; just why the MR did this is not clear, but afterwards the engine was not as good in performance as the other seven, and was the first one to be scrapped.

Eight engines were ordered from Beyer, Peacock & Co., intended to be numbered 87-94, but they were actually delivered in 1912 in Midland livery and numbered 2100-2107. Immediately the GER vetoed their use between Campbell Road Junction and Fenchurch Street, though they had previously raised no objections when informed of their axle loadings. This meant that on Fenchurch Street expresses, engines had to be changed at Barking if a 'Baltic' was in charge. There was no trouble about working into St Pancras. One good point from the operational view was that they did not have to be turned at the termini, since they had a symmetrical wheelbase, and could run in either direction with equal facility. With the 4-4-2 tanks on the fastest expresses it was always the rule that they had to run chimney first; with slower trains this was not enforced.

There were some teething troubles at first - inevitable with an entirely new design - but these were soon overcome, and the engines settled down to reg-

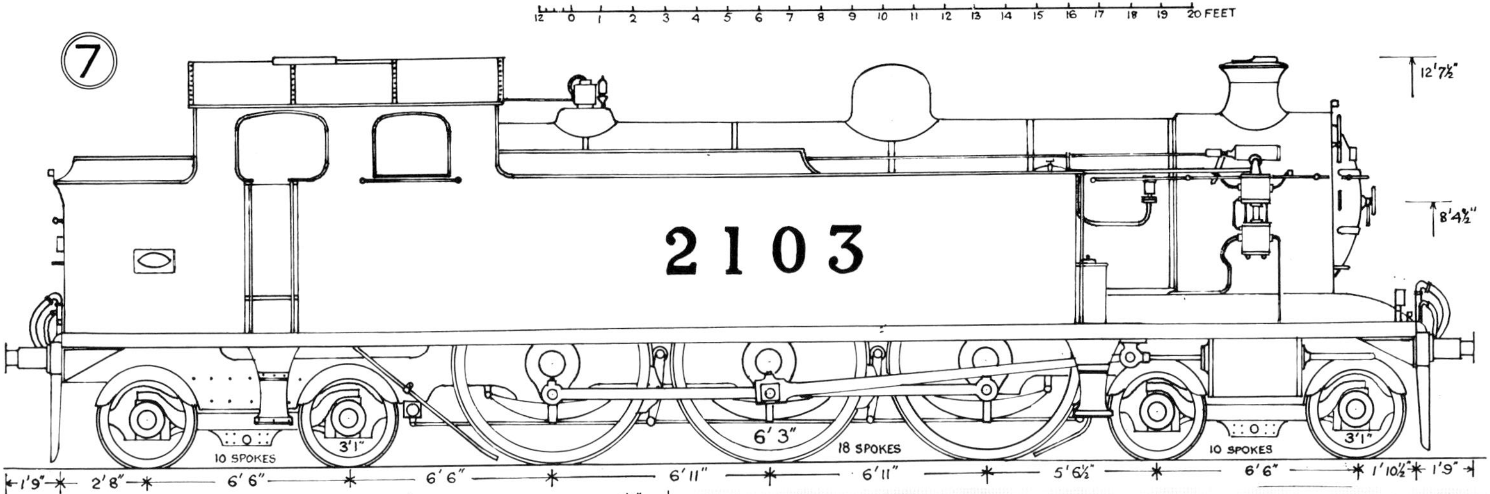

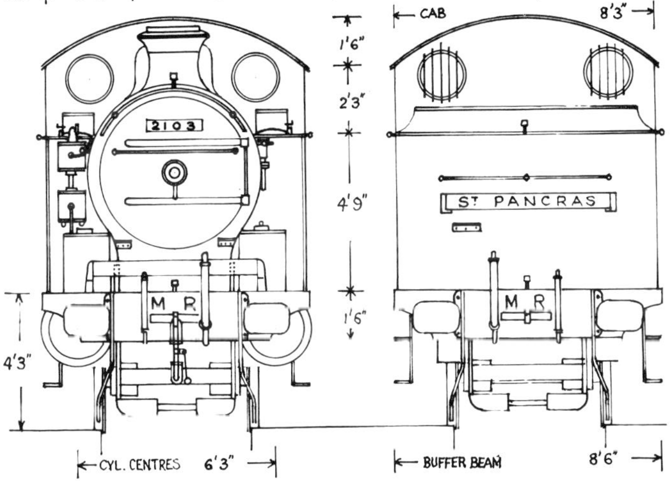

87 CLASS

8 BUILT 1912. BEYER, PEACOCK & Co.

ORDERED AS L.T.S.R. 87-94. DELIVERED AS MIDLAND RY. 2100-2107.

Plate 23: This is the second of the eight 'Baltic' tanks built by Beyer, Peacock in 1912. They came out with Midland Railway numbers; No. 2101 was later 2193. Note the dished smoke-box door, coal lid, and the MR coat of arms on the tank side in front of the cab entrance.

Real Photographs

Plate 24: No. 2100 at Plaistow in 1925, showing the Midland-type smokebox and chimney now fitted, and an LMS logo on the bunker with the maker's plate below.

H.C. Casserley

Plate 25: A view of '87' class 4-6-4T No. 2106 (ex-LT&SR No. 93). These were enormous locomotives weighing over 94 tons and built in 1912 by Beyer, Peacock & Co., Manchester. Withdrawal came in 1934, by then the locomotive carried the LMS No. 2198. It was the last of the class in service.

Real Photographs

ular work. They ran and steamed well, and were popular with the enginemen, all except No. 2100 after she had been 'de-superheated'. After the LMSR came into power, the 'Baltics' were transferred away from the Tilbury section, and some of them completed their career an coal trains from Toton and Wellingborough - a most unsuitable job for them.

The principal dimensions of the 4-6-4 tanks were - outside cylinders 20 in. x 26 in., coupled wheels 6ft 3in., bogie wheels 3 ft 1 in., wheelbase 6 ft 6 in. + 5 ft 6½ in. + 6 ft 11 in. + 6 ft 11 in. + 6 ft 6 in., total 38 ft 10½ in. The boiler was 5 ft 0 in. in diameter and 14 ft 6½ in. long, the outside firebox 8 ft 6 in. long. The number of tubes is not stated, but heating surface is given as 1305 (tubes) + 141 (firebox) + 319 (superheater), total 1765 sq. ft. Grate area was 25 sq. ft and working pressure 160 lb. The tanks held 2,200 gallons of water and 3 tons of coal were carried in the bunker. Weight in working order was 94 tons 4 cwt., with 53 tons 13 cwt. available for adhesion.

The first to be withdrawn was No. 2100, in February 1929. The rest were renumbered 2193-2199 in October of the same year, though No. 2107 never carried its new number (2199) as it was scrapped in September. Nos. 2101 and 2102 went in December 1929, followed by No. 2103 in October 1930. Thereafter, No. 2104 was withdrawn in December 1932, and No. 2105 in March 1933, leaving 2106 as the sole survivor; she was withdrawn in June 1934. It is a curious point that except for No. 2107, the class were withdrawn in numerical order; the author cannot recall any other class of locomotives treated in the same way.

None of the 'Baltics' were ever named. A scale model built in Plaistow Works to the original design was numbered 94 and given the name *Arthur Lewis Stride*, but this never appeared in full scale. It is not known whether it was the intention to name these eight engines; certainly the Midland was not keen on names, and removed all those of the LT&SR stock as they went through the paint shop.

Re-numberings

No.	*Makers*	*Date*	*1912*	*1923*	*1929*	*wdn.*
(87)	BP 5608	1912	2100	2100	--	2/29
(88)	BP 5609	1912	2101	2101	2193	12/29
(89)	BP 5610	1912	2102	2102	2194	12/29
(90)	BP 5611	1912	2103	2103	2195	10/30
(91)	BP 5612	1912	2104	2104	2196	12/32
(92)	BP 5613	1912	2105	2105	2197	3/33
(93)	BP 5614	1912	2106	2106	2198	6/34
(94)	BP 5615	1912	2107	2107	(2199)*	9/29

BP = Beyer, Peacock & Co., Manchester

* This number was allocated but never carried.

Livery

A few notes may be given here regarding the painting of LT&SR locomotives. The main colour was a bright green (No. 14 in Carter's British Railway Liveries) which was applied to boiler, tanks, bunker, cab, cylinders, footplate valance and wheels. Tanks, bunker and cab had an edging of purple brown with a fine red line inside it, these all having square corners. One inch inside these panels was a half inch black stripe edged on each side by a fine white line - these had rounded corners. The boiler bands were purple brown, with a red edging, and a black stripe edged with white appeared on each side of the boiler band. Buffer beams were vermilion, edged with black and a fine yellow line. Numbers (on buffer beams only) and names were in gold, shaded black. Wheels were green, with purple brown tyres. Main frames were purple brown, edged with black with a fine red line next to the black. Destination boards at front and rear were white, with block lettering in black.

The experimental livery applied to some of the rebuilt '37' class was described as lavender blue (not given in E.H. Carter's book *Britain's Railway Liveries*) which approximated to a bluish grey. The edging was dark blue, with white lining, and the inner stripe was crimson lake, also edged with fine white lines. Framing remained purple brown. This livery did not wear at all well, and was discontinued after 1910.

The four 4-4-2 tanks fitted with condensing gear for working in the District Railway tunnels were painted with Japan black; after 1905 Nos. 23 and 29 reverted to the standard green, only Nos. 3 and 7 remaining black. The two 0-6-0 tender engines, 49 and 50, were the standard green, and had the letters 'L.T.&S.R.' on the tender sides, in gold, shaded black. Square full stops were put after the letters (except '&') but about 1907 these were removed. After 1912, the Midland crimson lake was gradually applied to all passenger engines; those classed for goods service the 0-6-0s and 0-6-2 tanks, were painted black. Numbers were in standard Midland large figures on tanks or tenders, and a cast number plate was affixed to the upper part of the smokebox door. The Midland style of painting was continued under the LMS regime.

Plate 26: '87' class 4-6-4T No. 2107 (ex-LT&SR No.94) seen here at Derby. This locomotive had a short life being built in 1912 and was one of four members of the class which were withdrawn in 1929. *H.C. Casserley*

Plate 27: 0-6-0 No. 49 still looking an oddity as LMS No. 2898 in 1925, after the fitting of a Midland boiler. This engine, was one of only two 0-6-0s the LT&SR ever had. It was withdrawn in 1933. *H.C. Casserley*

Plate 28: '49' class 0-6-0 No. 2899 (ex-LT&SR No. 50) built by Sharp, Stewart & Co. in 1898, and seen here in August 1931, again with Midland boiler. *H.C. Casserley*

8

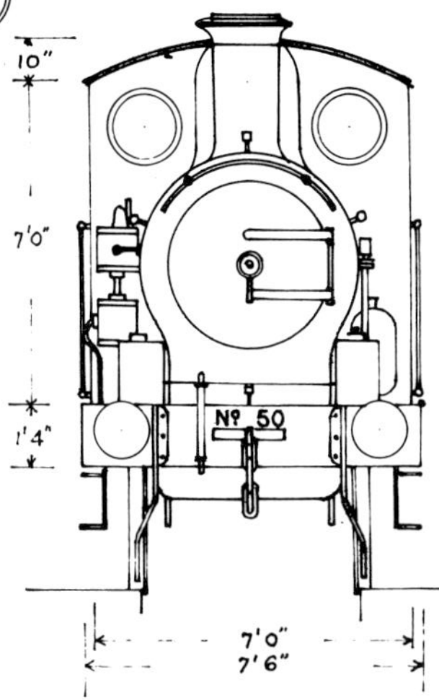

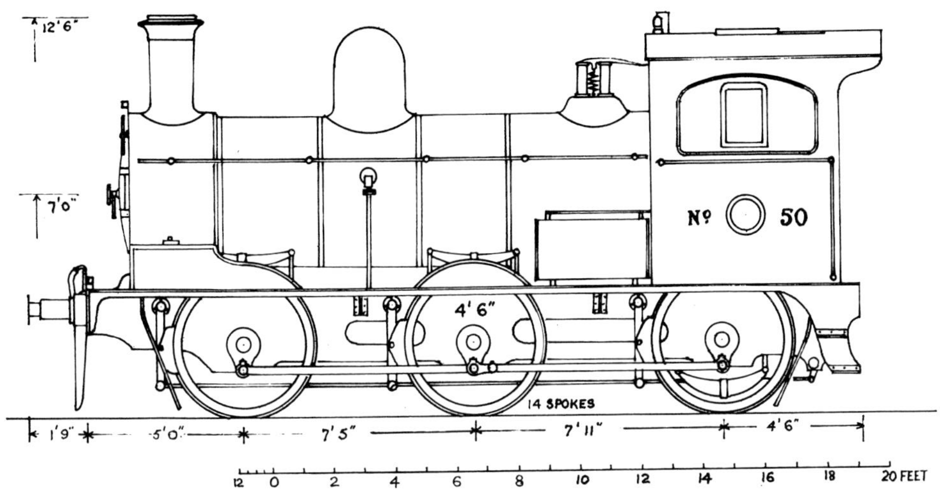

49 CLASS.

Nos. 49, 50.

M.R. & L.M.S. 2898, 2899.

BUILT SHARP, STEWART & CO.
(4420/1) 1898.
CYLS. 18" x 24"

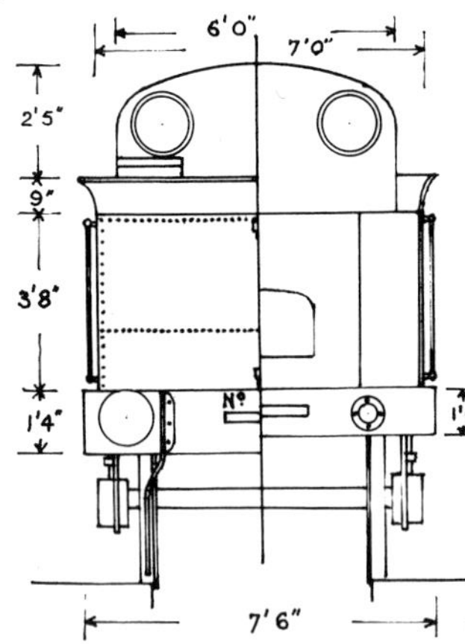

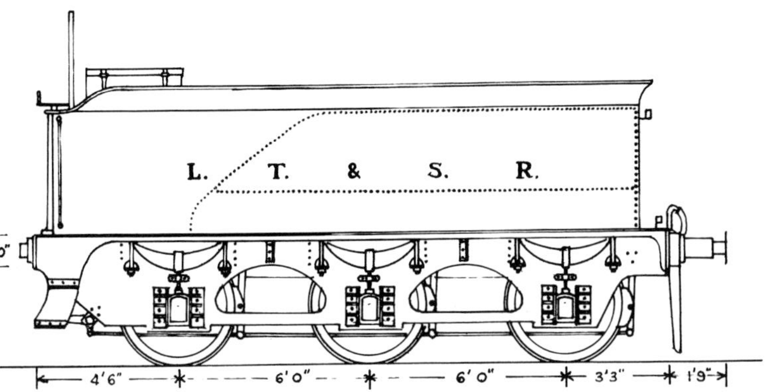

Locomotive Dimensions

	1 Class	37 Class	37 Class Rebuilt	49 Class	51 Class	69 Class	79 Class	87 Class
Type	4-4-2T	4-4-2T	4-4-2T	0-6-0	4-4-2T	0-6-2T	4-4-2T	4-6-4T
Wheels	3′1″, 6′1″, 3′1″	3′6″, 6′6″, 3′6″	3′6″, 6′6″, 3′6″	4′6½″	3′6″, 6′6″, 3′6″	5′3″, 3′6″	3′6″, 6′6″, 3′6″	3′1″, 6′3″, 3′1″
Cylinders	17″x26″	18″x26″	19″x26″	18″x24″	18″x26″	18″x26″	19″x26″	20″x26″
Wheelbase	6′6″+7′1″+ 8′6″+7′3″	7′0″+7′0½″+ 8′9″+8′0″	7′0′+7′0½″+ 8′9″+8′0″	7′5″+7′11″	7′0″+7′0½″+ 8′9″+8′0″	7′7″+9′3″+ 8′0″	7′0″+7′0½″+ 8′9″+8′0″	6′6″+5′6½″+6′11″ +6′11″+6′6″+6′6″
Wheelbase Total	29′4″	30′9½″	30′9½″	15′4″	30′9½″	24′10″	30′9½″	38′10½″
Boiler, Diameter	4′1″	4′2″	4′9″	4′4″	4′6″	4′6″	4′8″	5′0″
Boiler, Length	10′6″	10′6″	10′6″	10′0″	10′6″	10′6″	10′6″	14′6½″
Boiler, Pitch	7′1″	7′6″	8′3″	7′0″	7′9″	7′6″	8′3″	8′4½″
Boiler, Pressure	160	170	170	150	170	170	170	160
Tubes	200x1⅝″	201x1⅝″	212x2⅝″	216x1¾″	201x1⅝″	201x1⅝″	217x1⅝″	
Heating Surface Tubes	922	929	980	1018	929	929	1003	1155
Heating Surface Firebox	98	117	119	96	117	117	119	141
Heating Surface Superheater	-	-	-	-	-	-	-	319
Heating Surface Total	1020	1046	1099	1114	1046	1046	1122	1615
Firebox, Length	6′0″	6′9″	6′9″	4′10″	6′9″	6′9″	6′9″	8′6″
Grate Area	17.25	19.77	19.77	16	19.77	19.77	19.77	25
Water	1260	1500	1728	2640	1900	1914	1926	2200
Coal	2	2¼	2¼	3½	2½	2½	2¼	3
Weight, W.O.	56-2-0	63-0-0			68-3-0	64-13-3	69-7-2	94-4-0
Tractive Effort	13998	15552	17328	18288	15552	19254	17328	19930

Note There are at least three different versions of the Heating Surface of the 4-6-4Ts. The one given here is from O.S. Nock's *British Locomotives of the 20th Century Volume 1*. Likewise the number of tubes is also open to doubt.

Plate 29: Some early 4-wheeled compos, believed to be LT&SR Nos. 6, 12, and 1, at Upminster in 1926, now numbered LMS 03857, 03860, 03855. *H.C. Casserley*

Plate 30: A 48 ft third in British Railways livery at Bletchley in 1948. *H.C. Casserley*

Chapter Three

Passenger Rolling Stock

There is a degree of uncertainty concerning the LT&SR passenger stock, and a number of conflicting details have been handed down from different sources; thus it is difficult to decide what is most likely to be correct. The only definite information concerns those vehicles which were taken into stock by the Midland Railway in 1912, but here again there is no complete list of the MR numbers allotted to the ex-Tilbury coaches. A Diagram Book (of sorts) was issued by Plaistow Works, but this was sadly lacking in details, and the diagrams themselves were of a very sketchy nature. A document was unearthed many years ago by Mr A. Dunbar which purports to be a passenger diagram list of 1912. This is also very incomplete, and in itself produces several queries, since some of the dimensions given cannot be reconciled with definitely known details, while a few diagram numbers are blank, and in one or two cases there seem to be two different diagrarns for the same set of coaches. One wonders whether perhaps they were built by different makers, and therefore had separate diagrams while being otherwise the same. It is all very confusing, and whether it will ever be sorted out satisfactorily is anyone's guess. A copy of the Dunbar list is given in the Appendix, for what help it may be, but it seems to be the only available official document.

To complicate matters still further, the LT&SR was one of those companies which had a separate list, commencing at No. 1, for each class of carriage, plus several separate lists for non-passenger carrying stock (such as horse boxes, carriage trucks, milk vans, etc.). This method of numbering is said to have been recommended by the Railway Clearing House, their argument being that it assisted considerably with the RCH's work in assessing the charges due from one railway to another for through workings, etc. To the author's knowledge, no-one has yet succeeded in working out the logic of this statement. Be that as it may, the LT&SR was one of the companies which followed this dictum, and had separate lists for first class, second class, third class, brake third, and composite carriages.

Until second class was abolished on the Tilbury line in 1893 (though it did not disappear completely until 1904) the second and third class lists were apparently the same, for second class started at No. 1, and third class at No. 101. The gap between 16 and 101 was gradually closed, by both second and third class coaches. When second class was abolished, second/first composites became third/first; second/third became all third; and third/first remained unchanged. This process involved some renumbering, which complicated matters still further.

As the leasehold working of the line by the contractors expired in 1875, the LT&SR had then to find its own rolling stock, and a number coaches, all four-wheeled, were purchcased in 1876/7 from Brown, Marshall & Co. of Birmingham, and the Ashbury Co. of Manchester. These vehicles apparently sufficed until 1880, when the company acquired its own locomotives, and was no longer beholden to the Great Eastern Railway for working its trains. From then onwards, the LT&SR passenger stock expanded to keep pace with requirements, firstly with four-wheeled vehicles, and from the late 1880s six-wheeled vehicles; bogie stock was not introduced until 1901.

Also in 1901, in preparation for the through running with the Metropolitan District Railway, six nine-coach sets of four wheelers were purchased from Ashbury's designated as LT&SR/MDR Joint stock. These were numbered separately as M1-54. Each set was made up of one brake third, three thirds, two firsts, two seconds, and one brake second, and numbered in that order - M1-M9 in the first set, M10-M18 in the second set, and so on. Thus they did not conform to the general sequence of numbering. When the District Railway completed its electrification in 1905, these coaches were almost new, and the MDR share, Nos. M28-M54, were sold to the Taff Vale Railway in South Wales for excursion workings. The other 27 coaches were retained by the LT&SR for special work, and a 12-coach train of them was noted on an Ealing-Southend excursion in 1910. All except one of these 27 vehicles were reconstructed by putting two bodies together on new bogie underframes, 52 ft 9½ in. long, in 1912; five different types were thus produced. These were an eight compartment first; a 10 compartment and a nine compartment third; also two varieties of brake third, one with seven compartments and the other with six. There was some juggling with the original bodies, but not a great deal of actual 'surgery' to make them fit. This work is believed to have been carried out at Derby, but there is some doubt about this. The solitary brake second which was left over from these conversions is said to have been rebuilt as a full brake, but it cannot be traced in the stock list.

Much of the older four-wheeled stock was still in service with the Midland Railway until at least 1920, and some indeed survived long enough to be given numbers in the 1933 LMS renumbering list.

All early coaches were four-wheeled, with a common width of eight feet over waist, but length and wheelbase varied according to types and makers. Most of them were built by either Ashbury's of Manchester or the Metropolitan Railway Carriage & Wagon Co. of Saltley, Birmingham. The earliest carriages were lit by oil lamps, one to each compartment; oil gas was used for a short period in the 1880s and early 1890s, but in 1896 the company decided to change over to electric lighting. New stock from that date was fitted with electric lighting when built, and all other stock was converted gradually. The LT&SR was one of only two steam-hauled railways to have its entire passenger stock lighted by electricity.

General features of the coaches were as follows: low arc roofs, ends straight, divided into seven vertical panels by narrow half-round beading; sides straight, with a small amount of tumblehome at the bottom, panelled by half round beading with square corners to each panel, doors panelled separately. Cant rail panels and the louvred ventilators in the tops of the doors had semi-circular ends, except in the Joint stock of 1901, which had semi-circular louvred ventilators of the peculiar design produced by Ashbury. Windows had curved upper corners and square lower corners. There were four footsteps for roof access with curved handrails, fitted at one end only; the opposite end carried the lighting switch. Indicators for the communication cord were fitted at both ends. A full length footboard was bracketed to the botton of the solebar on each side, with a second footboard at axle level. Axleguards were standard 'W' irons, with long springs. Westinghouse brake gear was fitted from the mid-1880s.

Livery was varnished teak, with gold lining on the middle of the beading, but not applied to the end panels. Lettering was 'L.T.&S.R.', with full stops, in gold, shaded red on the waist panels. Class indication was originally in words on the door waist panels, but from 1908 this was changed to a large number on the lower door panels. Torpedo ventilators were fitted to the roof of smoking compartments only; from 1908 'SMOKING' also appeared on the door waist panels of the relevant compartments. Underframes, bogies and other ironwork were painted black; wheel tyres were painted white, though this was not universal. Roofs were light grey, though this was not always apparent unless the coach was newly out of workshops. Except for minor details, these features applied also to the six- and eight-wheeled stock. The only major difference, begun with the six wheelers, was a change in roof profile to what is usually known as a 'cove' roof - deeper but slightly flatter, with a sharp curve down into the eaves - very similar to the Great Northern standard, and also used on a considerable proportion of London North Western stock.

Details of each type of coach follow, beginning with the four-wheeled varieties; the Diagram numbers are also given, where known, and the number of the relevant line drawing. Not all the individual types of coach are illustrated; for one or two of them no illustration seems to be available, and for some, where the only difference is in a matter of an inch or two in length, one drawing can serve for both. Running numbers are given as far as possible; there seems to have been a fair amount of renumbering, not much of which is clear, and the numbers given are those appertaining in 1912. That renumbering has taken place at some stage or other is apparent from the numerous instances of building dates out of sequence.

Plate 31: A view taken on the Corringham Light Railway on 17th May, 1947. The coach is ex-LT&SR third No. 11 which was built 1876/7. *H.C. Casserley*

Plate 32: Midland Railway Co. plate on LT&SR 'Ashbury' coach No. 23243 photographed on 23rd August, 1955. *H.C. Casserley*

Plate 33: Seen at Thames Haven on 10th January, 1953 is the 12.20 pm workers train to Tilbury. Coaches are No. 47458 and 23245 (E14040M). *H.C. Casserley*

FOUR-WHEELED STOCK

In 1876/7 the first coaches were delivered to the company, the exact number not being immediately apparent, due to the renumbering above mentioned. They varied in length, the first class being 26 ft 0 in. in length, the seconds 28 ft 4 in., and the composites 24 ft 9 in. Seating was for six in first class compartments and 10 in seconds and thirds. Second and third class coaches had five compartments, and firsts and composites four compartments. All had two smoking compartments, in the case of first class one at each end, and in the others two compartments at the same end. Wheelbases varied between 14 and 16 feet. Further coaches of these types were added to stock at intervals between 1880 and 1886. In 1893 second class compartments were downgraded to thirds, and some coaches *may* have been renumbered into the third class list, but more probably still retained their original numbers. These three types of carriages are shown in *figure 9, 10 and 11*; the Diagram numbers were probably 2 (first class), 12 (second class) and 7 (composite). Some of the first class may have been on diagram 1, as they had a wheelbase of 15 ft 3 in., but this is not confirmed.

Some of these early coaches were taken into Midland Railway stock in 1912, and received MR numbers 2444 upwards, being put on the Duplicate List (with 0 in front of their numbers) from 1914 onwards. Nos. 02444-02459 were replaced in running stock in 1914, and Nos. 02460-02474 in 1921. Three ex-LTS thirds (formerly seconds) Nos. 2, 7, and 9, were renumbered MR 02445/49/51, and were sold to the Corringham Light Railway in 1915.

The first of the third class coaches are said to have been Nos. 101-124, received in 1883, of Diagram 12, and duplicates, except for class, of the earlier seconds, having a length of 28 ft 4 in., with five compartments. The generally accepted version is that Nos. 1-100 were reserved for second class coaches, and 101 upwards for third class. However, although all the numbers up to 100 had been filled by 1883, it is by no means certain that all of these were second class; it seems very likely that at least some of them were delivered as third class. Certainly, by 1900, all of them were third class, and with the abolishing of the second class list entirely, there is no means of finding out. Until 1883, the guards were accommodated in full brake vans (Diagram 29) of which there were 17, Nos. 1-7 delivered in 1877, Nos. 8-10 in 1883, and Nos. 11-17 in 1886 (Fig. 15). There may have been others, since one Joint stock brake second is known to have been converted to a full brake in 1912, and there are rumours that some older passenger coaches were converted as well, but these cannot be found in the stock list, and must be taken as doubtful.

These full brake vans were 26 ft long, and had lookout duckets in the centre, with the guard's door adjoining; each half of the van was provided with a pair of double doors on each side for dealing with large objects. The

Plate 34: This early four-wheeled third carriage had half-height partitions except for the centre compartment; this is No. 23, believed built in 1886. *Real Photographs*

Plate 35: Centre brake No. 51, after renumbering in the third class series, was built in 1891. Note the safety chains fitted at the time on all carriages. *Real Photographs*

9

DIAG. 12.

THIRD CLASS.
25'9" × 8'0"

SECOND CLASS.
28'4" × 8'0"
TO ALL-THIRD, 1893

Nos. 101-116 (1880). 117-124 (1882) 23-28 (1886) 29-32 (1887) 39-50 (1890)
TOTAL 58.
SECOND CLASS Nos. 1-12 (1880) 13-16 (1882)

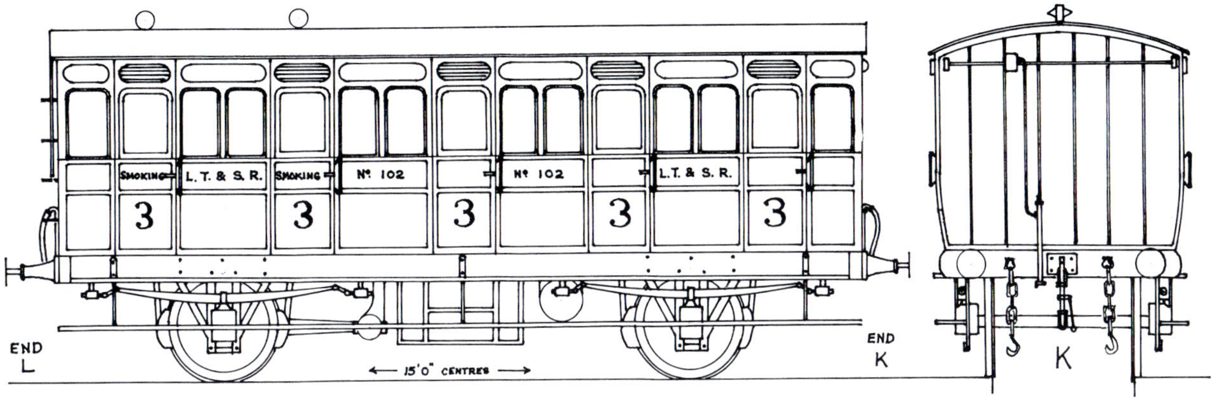

10

DIAG. 2.

FIRST CLASS.
26'0" × 8'0"

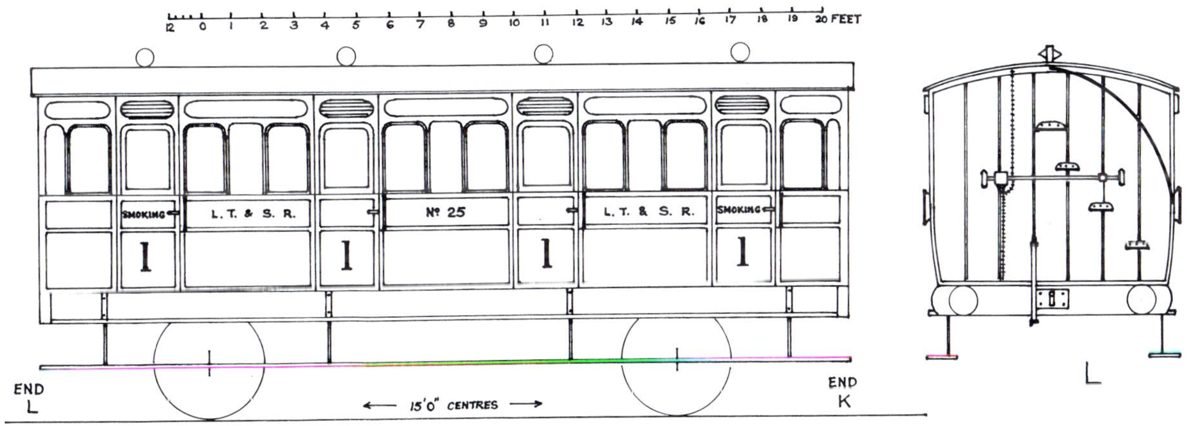

Nos. 1-12 (1880). 13-16 (1882) 25-36 (1886)

Plate 36: Third brake No. 1, built about 1879, is fitted with spoked wheels in place of the more usual Mansell type. *Real Photographs*

(11)

Nos. 1-6 (1876). 7-12 (1880)

DIAG. 7

COMPOSITE.
SECOND/THIRD.

24'9" x 8'0".

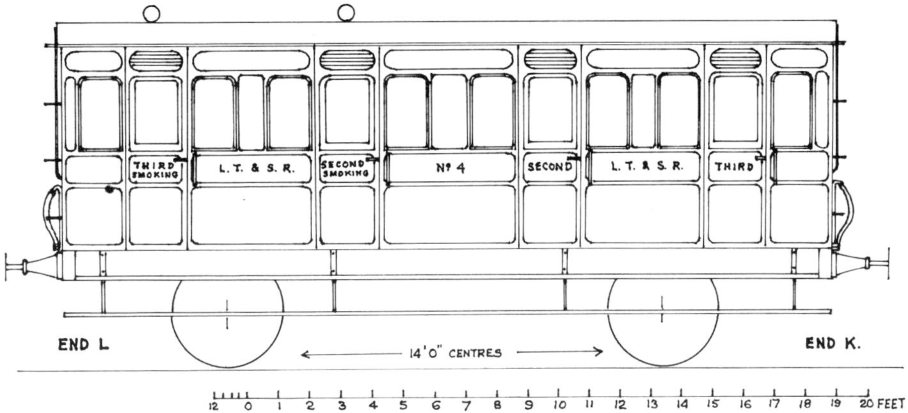

For end views K and L refer to page 53.

(12)

Nos. 13-18 (1883)

DIAG. 20

BRAKE THIRD.

24'9" x 8'0"

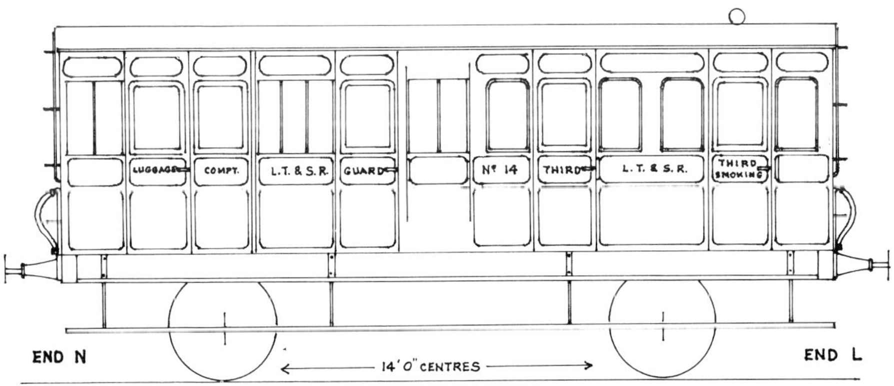

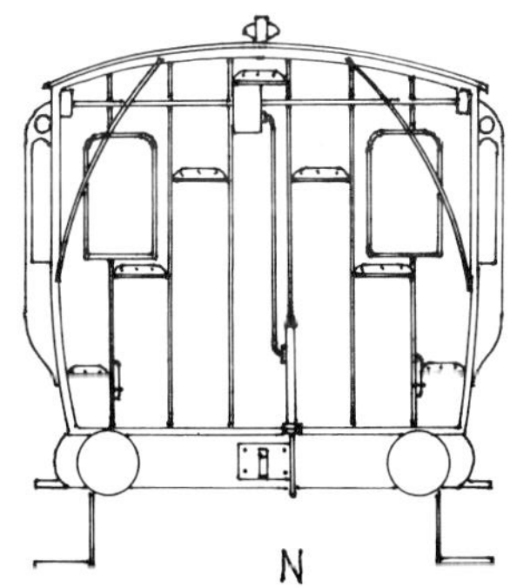

extreme width over duckets was 9 ft, and this became the standard dimension for all this type of vehicle.

There were 13 brake third coaches with two compartments which dated from 1877/8, numbered 1-13, and having the guard's van at one end. Two diagrams applied, Nos. 1-9 being on Diagram 19, and Nos. 10-13 on Diagram 20. The only difference was that Diagram 19 vehicles were 25 ft 9 in. long, and those based on Diagram 20 one foot shorter. These vehicles are alleged to have been originally brake seconds, but this cannot be confirmed. Fig. 12 illustrates them. The advent of Nos. 14-21 in 1883 settled the matter of whether to have full brake vans or not, since no more full brakes were built, and brake thirds became the rule. These eight coaches were of a different type; a compromise with the full brakes, since the guard's van was in the centre, and there was one third class compartment at each end. A new Diagram 21, was issued, to cover these, and they were added to in two further batches, Nos. 22-28 in 1884 and Nos. 43 and 44 later in the same year. In 1886 Nos. 51-54 appeared, and four more, Nos. 29-32 were also added to stock, though the date in the stock list - 1891 - seems rather doubtful. In all these 25 vehicles, the central duckets were as before, but there was no separate guard's door, this being incorporated as one of the double doors. In most of these brake thirds, the guard's door was arranged to open inwards. The Midland renumbering of the Diagram 21 coaches is a little suspect; basically they were given numbers 4346-4365, but 4350 was omitted and an odd number - 3892 - was inserted. Four of them do not appear to have received a Midland number; full details are given in the stock list.

Two odd four-wheeled coaches, first class No. 17 and third class No. 136, appear to have been renumbered, though what their original numbers were is impossible to say. Both were short vehicles, only 24 ft 8 in. long. No. 17 was built in 1883, and was on Diagram 3, while No. 136 was on Diagram 15, and was dated 1877. In both cases, there was only the single coach on the diagram. No. 136 replaced a six-wheeled third scrapped after the Pitsea accident in 1897, which gives a clue to the date of renumbering. No. 17 became Midland 2798 in 1912, and No. 136 was MR 4250. Neither of them survived into LMS stock.

There remain now the Joint stock, built in 1901 by Ashbury. These were in five types, brake third, brake second, first class, second class, and third class. All had a common length of 26 ft 5in., and wheelbase of 15 feet; they were close-coupled, with short buffers, though the outer ends of the brakes were fitted with normal buffers. They all had the typical Ashbury features, semicircular louvre ventilators in the top of the doors, square cornered windows and panels, with the waist panels of equal depth to the lower panels. The doors were panelled separately. Roof ventilators of the smoking compartments were not of the torpedo pattern; they were rather like two small flower

13 DIAG. 3.

FIRST CLASS.

DIAG. 3

24'8" x 8'0"

Nº 17 ONLY.

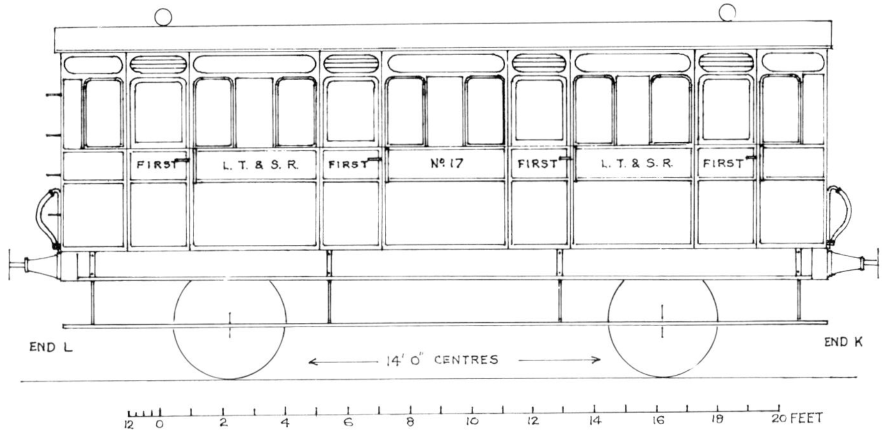

For end views K and L refer to page 53.

14 DIAG. 15

THIRD CLASS

DIAG. 15

24'8" x 8'0"

No. 136 ONLY

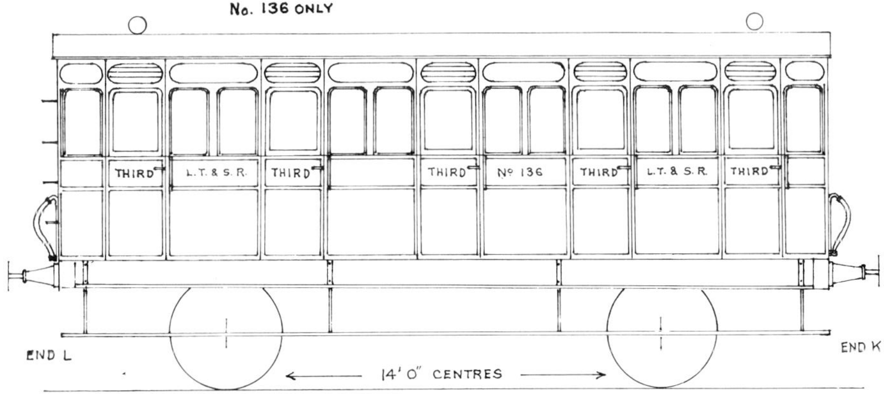

15

DIAG. 29

BRAKE VAN

26'0" × 8'0"

Nos. 1-6 (1880) 7-12 (1882) TOTAL 12.

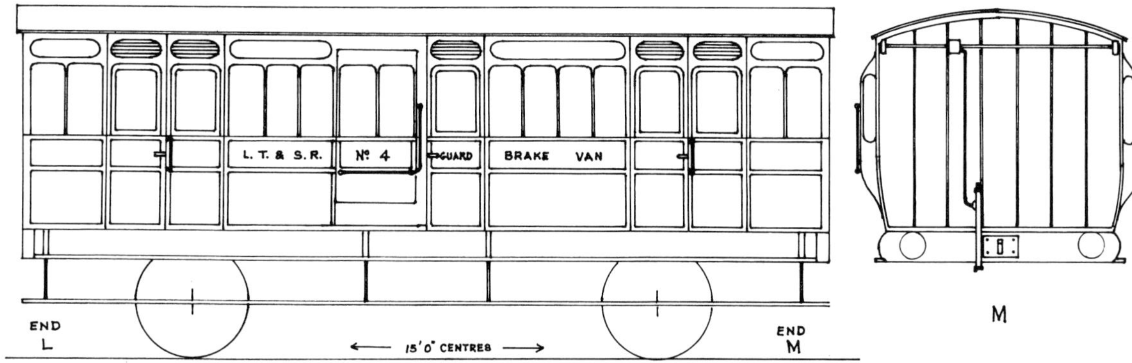

16

DIAG. 21

BRAKE THIRD

25'9" × 8'0"

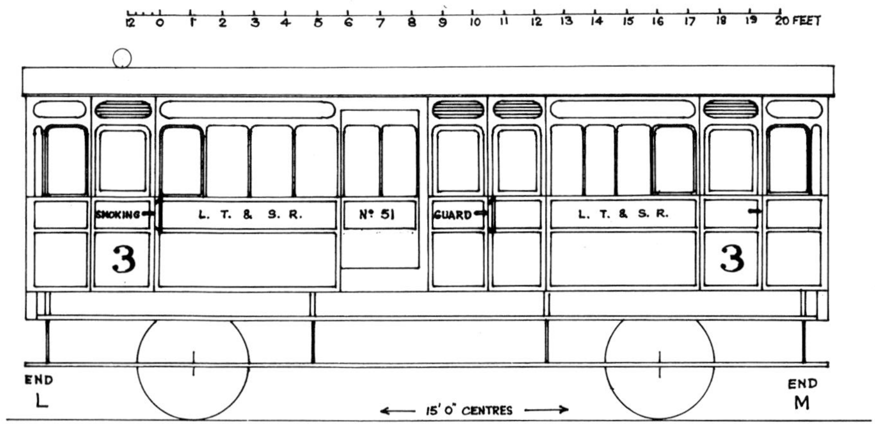

For end view L refer to page 53.

Nos. 51-53 (1891)

Plate 37: First/third class composite No. 6 (1879) had very wide first class compartments. *Real Photographs*

Plate 38: This coach numbered 50 was clearly designed as a 1st/2nd composite but appears here as all-first; body length was 26 ft 9 in. *Real Photographs*

17 THIRD CLASS.

DIAG. 13. 14.

DIAG. 13 28'3" x 8'0"

DIAG. 14 28'4½" x 8'0"

DIAG 13. Nos. 52. 53. 55. 60. 61. 66 (1878). 182. 186. 187. (1881) TOTAL 9.

DIAG. 14. Nos. 173-181, 183, 184. (1880) TOTAL 11.

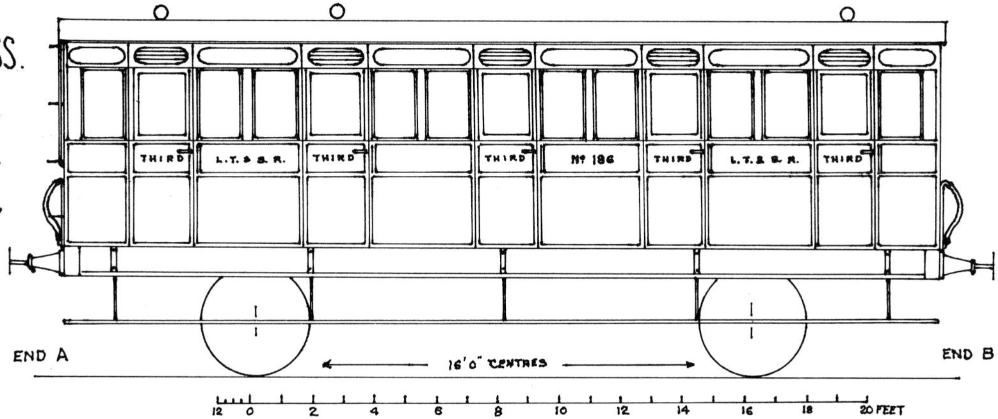

For end views A and B refer to page 85.

18 FIRST CLASS.

DIAG. 4

30'0" x 8'0"

Nos. 37-44 (1893) 52-54 (1899) TOTAL 11.

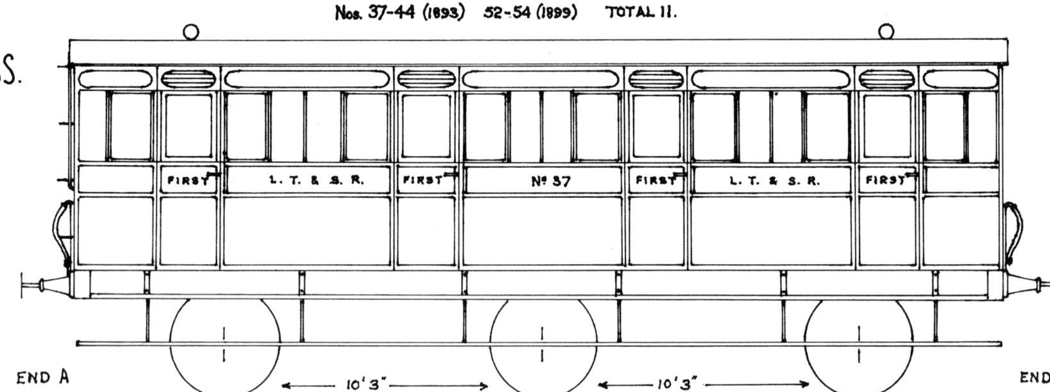

(19) JOINT STOCK. 1901. Nos. M1. M10. M19. TOTAL 3.

BRAKE THIRD.

26'5" x 8'0"

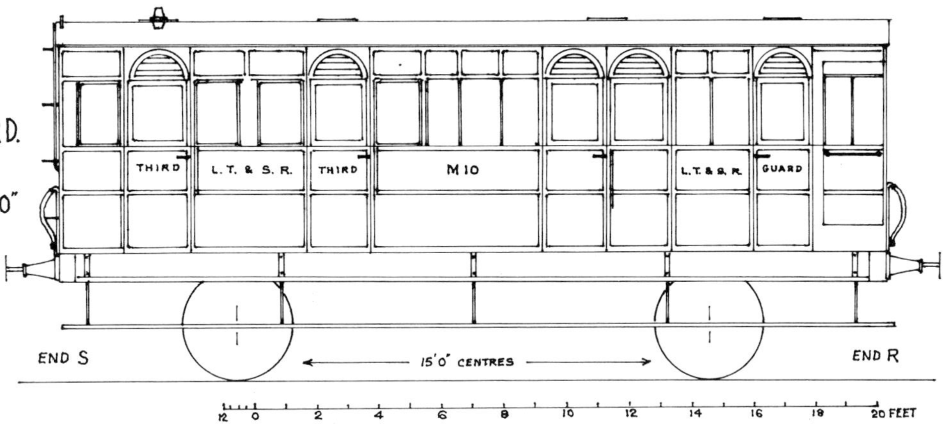

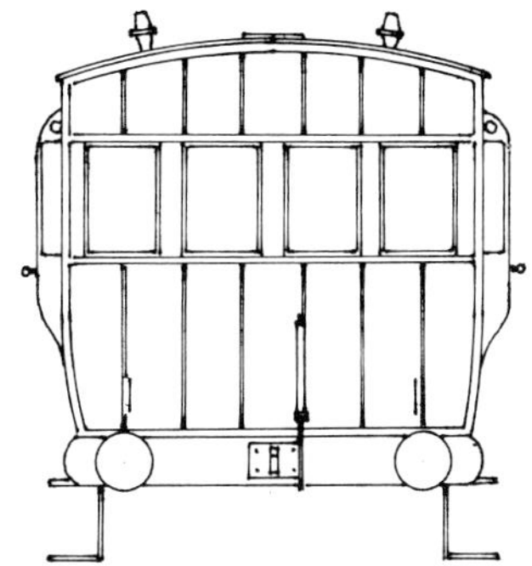

(20) JOINT STOCK. 1901. Nos. M2. M3. M4. M11. M12. M13. M20. M21. M22. TOTAL 9.

THIRD CLASS.

26'5" x 8'0"

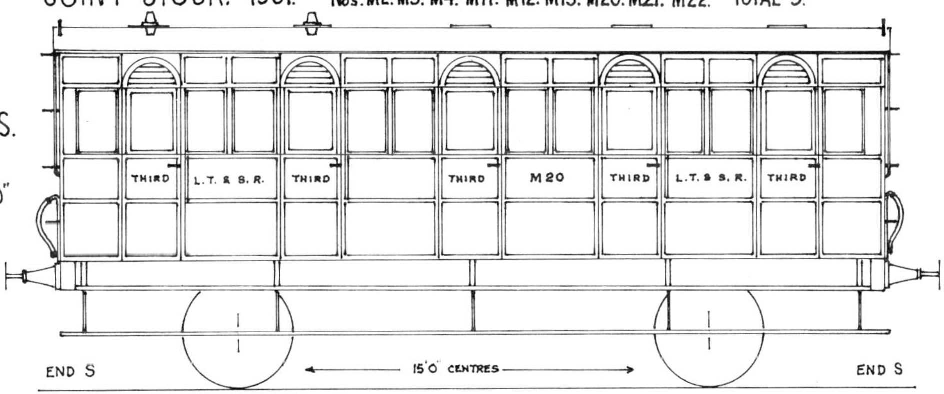

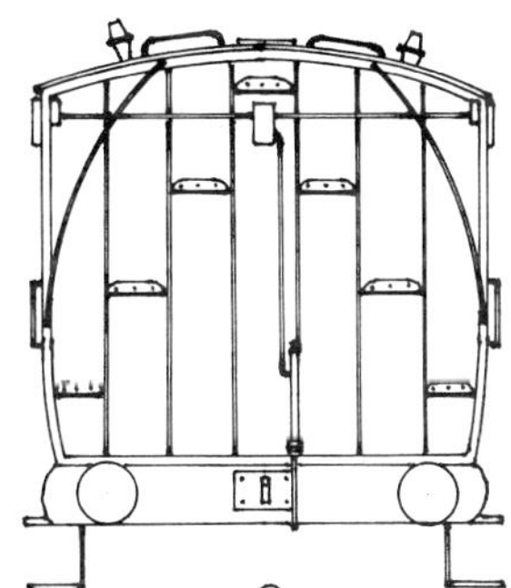

ALSO SECOND CLASS Nos. M7. M8. M16. M17. M25. M26. TOTAL 6.

21

JOINT STOCK.
1901

BRAKE SECOND.

26'5" x 8'0"

Nos. M9. M18. M27. TOTAL 3.

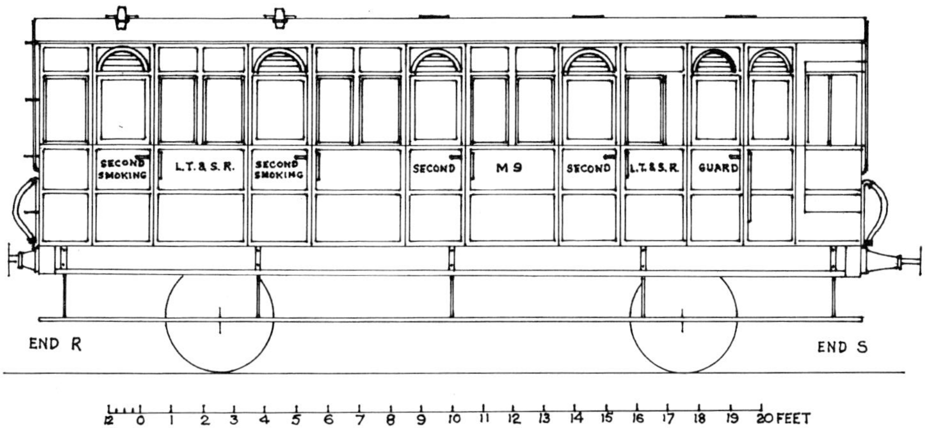

For end views R and S refer to page 61.

22

JOINT STOCK.
1901.

FIRST CLASS.

26'5" x 8'0"

Nos. M5. M6. M14. M15. M23. M24. TOTAL 6.

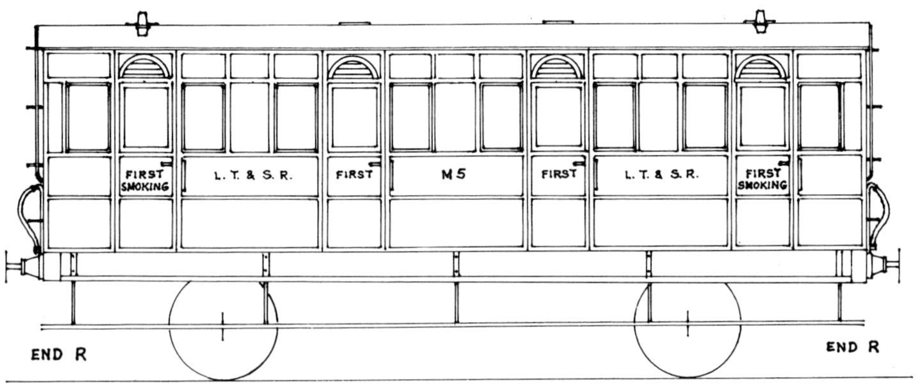

pots placed rim to rim. Access steps to the roof were fitted to both ends, but not to the van ends. Cant rails were rather unusual in that they had a small panel over each window instead of one long panel spanning both adjacent windows. Second and third class coaches had five compartments, two smoking at one end; first class had four compartrnents, with one for smokers at each end. The brake thirds had two compartments, at one end, with the guard's lookout duckets at the other end, a separate guard's door, and a pair of double doors in the van. Brake seconds had four compartments, with a very short van at one end. It has not been thought necessary to illustrate the second class coaches, since they were exactly the same as the thirds (*figures 19 & 22*).

The entirely separate numbering of this stock has been mentioned earlier. Each nine coach set was made up in the same order, brake third, three thirds, two firsts, two seconds, brake second, and numbered in that order, beginning with the brake third M1. So the three sets retained by the LT&SR in 1905 carried the numbers M1, M10, M19 (brake thirds), M2, M3, M4, M11, M12, M13, M20, M21, M22 (thirds), M5, M6, M14, M15, M23, M24 (firsts), M7, M8, M16, M17, M25, M26 (seconds), M9, M18, M27 (brake seconds). It might be wondered why, since second class was abolished on the LT&SR in 1893, the Joint stock still had second class coaches. The reason for this was that the Metropolitan District Railway still had second class coaches, so the Joint stock was built to conform.

All the Joint stock coaches were rebuilt by putting two bodies onto new 52 ft 9½ in. bogie underframes in 1912, leaving only one brake second unused. This is reputed to have been converted to a full brake, but if so, it can not be traced in the stock list; there is a possibility that it was transferred to departmental use. The rebuilt bogie coaches will be dealt with in due course.

There were three saloon coaches in the stock list, two four-wheeled (Diagram 27) and one six-wheeled (Diagram 28). The four-wheeled saloons were 24 ft 9 in. long, and the six-wheeler 34 ft 0 in. in length. No illustration of either type is available, nor any details of their internal arrangements, or passenger classification. The only information available is that the two four-wheeled coaches were numbered 1 and 2 (in a separate series) and the six-wheeled saloon was No. 3. The Midland Railway allotted Nos. 4366/7 and 2800 to them; none of them appear in the LMS 1933 list.

Six Wheeled Stock

Though the six-wheeled coaches outnumbered considerably the four-wheeled stock, there were only six relevant diagrams. These vehicles were enlarged versions of the earlier types, and had a common wheelbase of 20 ft, but varied in length according to type. With these coaches came the 'cove' roof, which was to remain the standard pattern for the rest of the stock. There were two varieties of composites built, variant in length, but otherwise the same in most details. Diagram 8 vehicles were 32 ft 0 in. long, with a third class compartment at each end, and three first class in between; the adjoining third and first at one end were designated for smokers. The other type, Diagram 9, showed coaches 29 ft 9½ in. long, with three second class compartments instead of flrst. In all there were eight vehicles on these two diagrams, presumably four of each and numbered 17-24. *Figures 23 and 24* show these two types.

In 1912 they became numbers 3843-3850 in the Midland list, but only one, (24) was eventually given a number - 27210 - in the 1933 LMS stock.

Eleven all-first class coaches were built, also 29 ft 9½ in. long, in two batches, Nos. 37-44 in 1893 and Nos. 52-54 in 1899 (Diagram 4, shown in *figure 18*). There were four compartments, one at each end being for smokers. These became numbers 2495-2600, 2602-6 in the Midland list, but were not included in the LMS renumbering. Only six brake thirds were built, 34 ft 0 in. long, numbered 33-38; these appeared in 1888, were Diagram 22, and are shown in *figure 26*. These had three compartments, two at one end and one at the other, with the guard's van in the middle. The lookout duckets adjoined the inner of the two compartments, with the guard's door next; the usual double doors were approximately in the middle of the van section. Their Midland numbers were 4318- 4323 and four of them lasted long enough to be given Nos. 27754-27757 in the 1933 LMS list; the two scrapped early were former Nos. 33 and 37.

Four further brake thirds were added to stock in 1899, Nos. 39-42; this batch were 31 ft 6 in. long, and had only two compartments, one at each end. These were to Diagram 23, and are shown in *figure 27*. In 1912 they were renumbered MR 4324-4327, only one (39) surviving long enough to become LMS 27774.

By far the largest class numerically was the six-compartment thirds of Diagram 17 (*figure 25*) which were 34 ft 0 in. long, and ultimately numbered 79, built over a period of 11 years. The first of the class were built in 1888, nos. 125-136, and thereafter batches came out at irregular intervals. The first 12 were followed by 12 more in 1890, Nos. 137-148, then 149-164 in 1891, and 165-172 in 1892. After these came a gap of six years, until 1898, when Nos. 188-203 were built; Nos. 204-218 followed in 1899. The whole of the six-wheeled thirds (with the exception of the three scrapped after the Pitsea accident in 1897, Nos. 136, l43 and 168) were renumbered into Midland stock as 2342-2417, but not necessarily in order; 55 appeared in the LMS 1933 list as 26524-26579.

23

DIAG. 8

COMPOSITE FIRST/THIRD.

32'0" x 8'0"

Nos. 17-20 (1891) TOTAL 4.

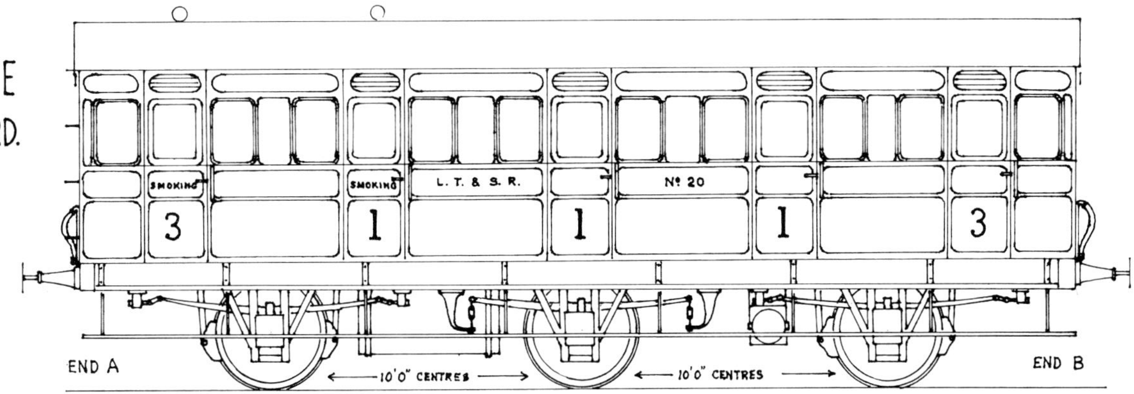

For end views A and B refer to page 85.

24

DIAG. 9

COMPOSITE SECOND/THIRD

29'9½" x 8'0"

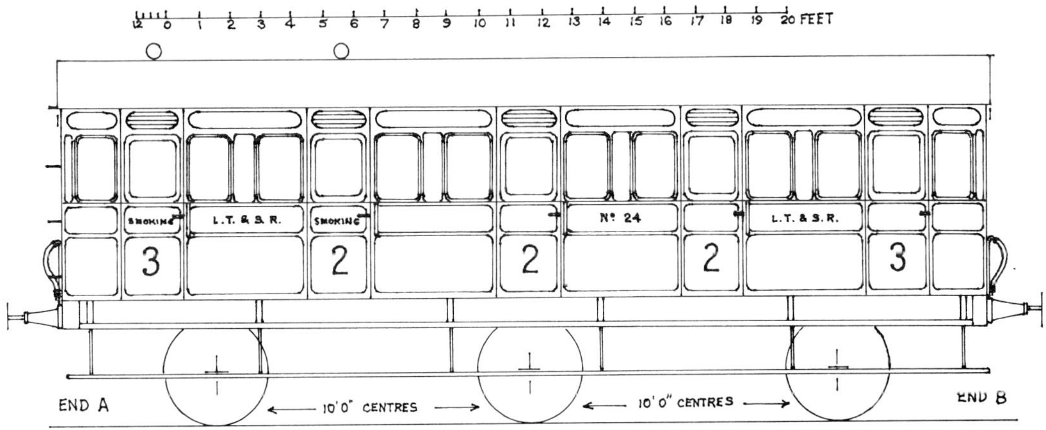

Nos. 21-24 (1891) TOTAL 4.

Nos. 125-136 (1889) 137-148 (1890) 149-164 (1891) 165-172 (1892) 188-203 (1898) 204-218 (1899)
PLUS 173-184 (1891) DOWNRATED FROM SECOND CLASS IN 1893. TOTAL 91

25 DIAG. 17.

THIRD CLASS.

34'0" x 8'0"

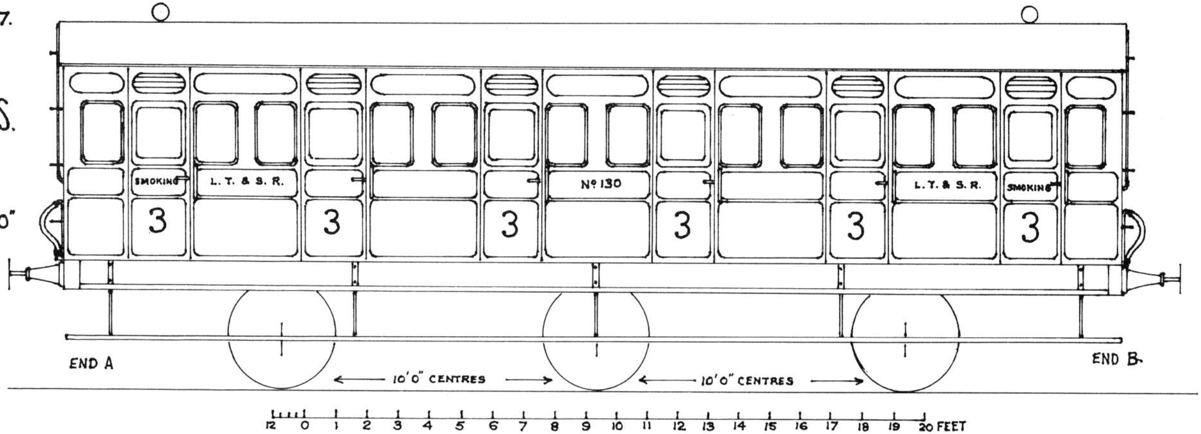

26 DIAG. 22

BRAKE THIRD.

34'0" x 8'0"

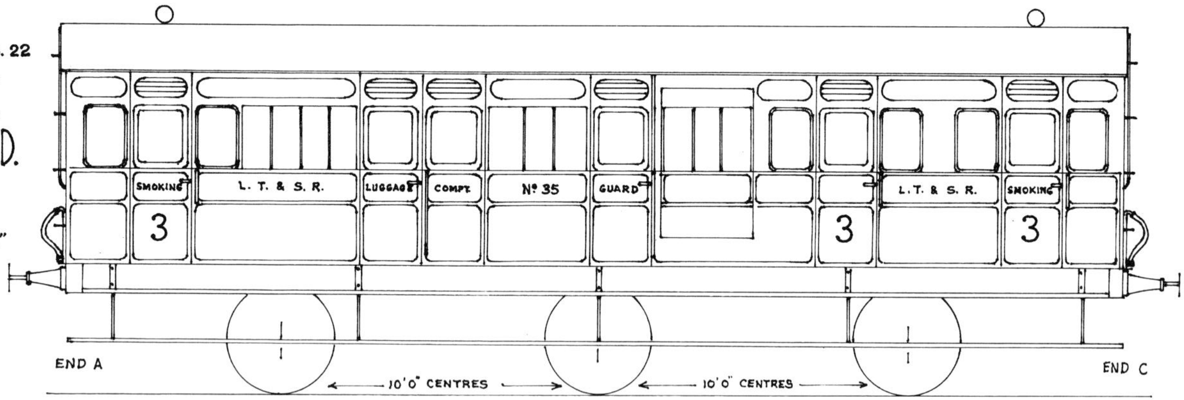

Nos. 33-38 (1889) TOTAL 6

For end views A, B and C refer to page 85

27

BRAKE THIRD

DIAG. 23

31'6" x 8'0"

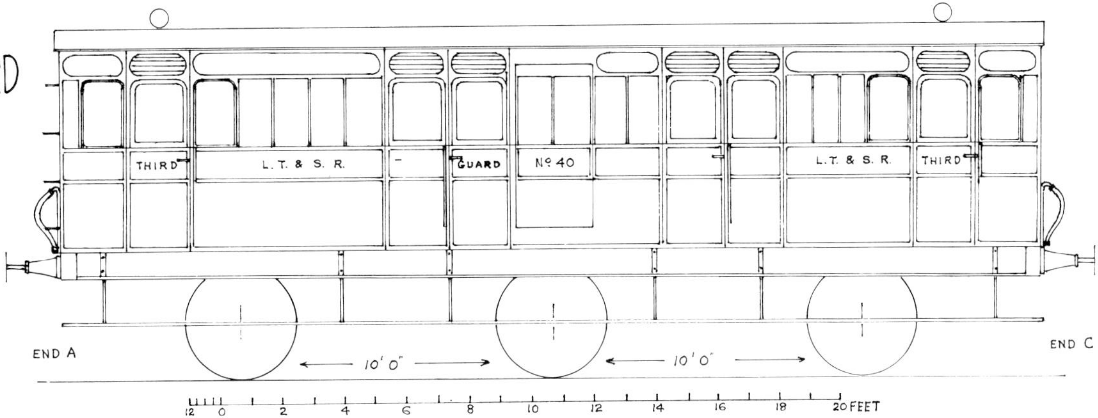

28

For end views A, C, F and G refer to page 85.

BRAKE THIRD.

DIAG. 25

46'0" x 8'0"

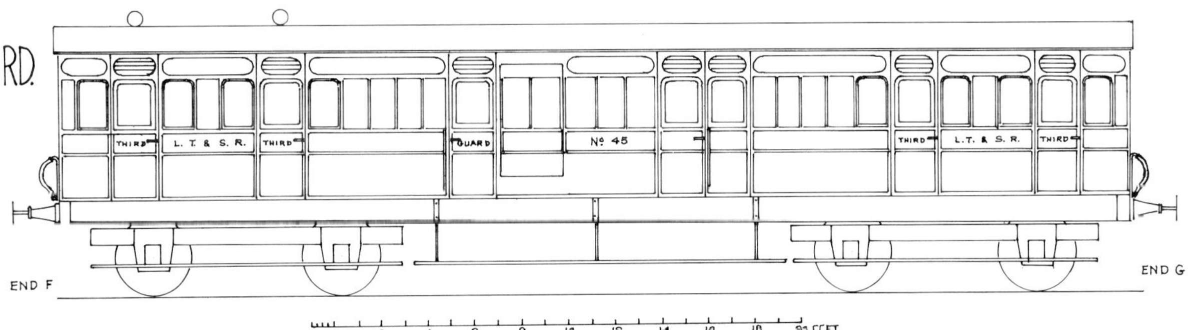

Eight Wheeled Stock

In 1901 bogie carriages were introduced. These were of a standard length of 48 ft for first class and composites, and 46 ft for third class and brake thirds. Why the thirds were made two feet shorter is a mystery, since they would have benefitted by the extra length - they were rather cramped in compartment length. It may be that the length of some station platforms had something to do with it, Fenchurch Street in particular. The standard width was 8 ft as before, but in 1908 this was increased to 8 ft 4 in., without any alteration to the seating capacity. Two 11-coach trains were built in 1901, consisting of one composite, two firsts and eight thirds in each; for the time being no brake thirds were provided, the older six-wheeled type being used. The thirds had eight compartments, the firsts seven, and the composites six thirds and two firsts (in the centre). In 1904 an experiment began of fitting lavatories to some first class and third class coaches; this was done by blanking off one half of two adjoining compartments, producing two lavatories of very cramped dimensions, and serving only two half-compartments. The windows were fitted with frosted glass, and a long louvre ventilator fitted into the cant rail. Mostly the lavatories were fitted in the centre of the coach, but not always, as there is evidence of them being fitted elsewhere. Only one first class coach (No. 60) was so fitted, but five thirds were so altered, Nos. 222, 225, 229, 233 and 234. Later batches of these eight wheeled coaches also had odd ones fitted with lavatories, but in all only three firsts and eight thirds were eventually so fitted. None of the brake thirds or the composites ever had lavatories.

Additional batches were built in 1904, 1906, 1908, 1910 and 1911, until there were a total of 20 first class, 21 composites, 21 brake thirds, and 73 thirds. The first four brake thirds were built later in 1901; these had centre guard's vans with two compartments at each end, one of which was for smokers. Two more were built in 1904, but when further brake thirds were required in 1906, the design was altered, the van being placed at one end and the four compartments together at the opposite end; the van portion had the usual duckets and double doors.

These standard bogie vehicles were shown on five diagrams, and are illustrated in these pages by *figures 28 to 32*. No separate Diagrams were issued for the lavatory-fitted variants. The first class coaches were on Diagram 6, the composites on Diagram 10, third class on Diagram 18, while two diagrams, 25 and 26, covered the two types of brake thirds.

The full numerical list of these bogie carriages, together with the Midland and LMS renumberings, is given in the stock list in the Appendices.

Towards the end of 1911, two eight-coach trains with corridor facilities were ordered by the LT&SR for the joint through service with the District

29

DIAG. 26

BRAKE THIRD

46'0" × 8'0"

Nos. 69-71 (1908) 72-75 (1910) 76-79 (1911) TOTAL 11.

For end views F and G refer to page 85.

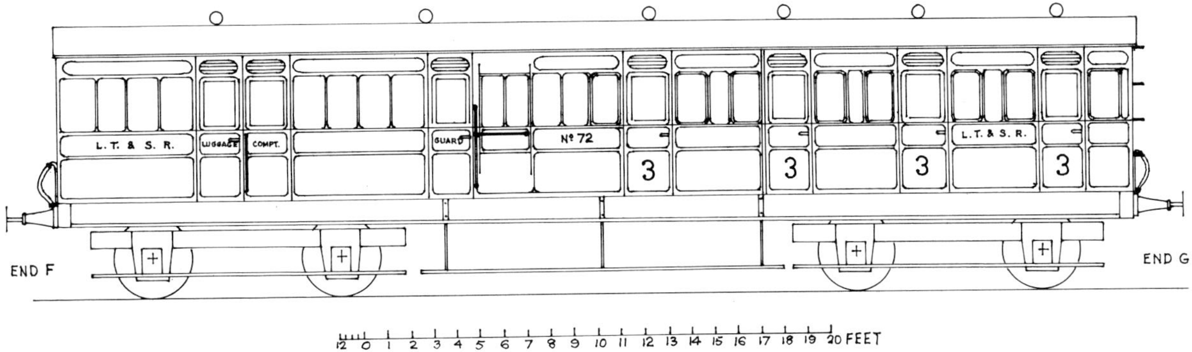

Plate 39: Devons Road carriage sheds in 1925 showing LT&SR six wheeled carriage stock on the left. *R.W. Kidner*

30

DIAG. 10

COMPOSITE.

Nos. 25.26 (1901) 27 (1904) 28-32 (1906) 33-36 (1908) 37-41 (1910) 42-45 (1911) TOTAL 21.

For end views A and B refer to page 85.

48'0" × 8'0"

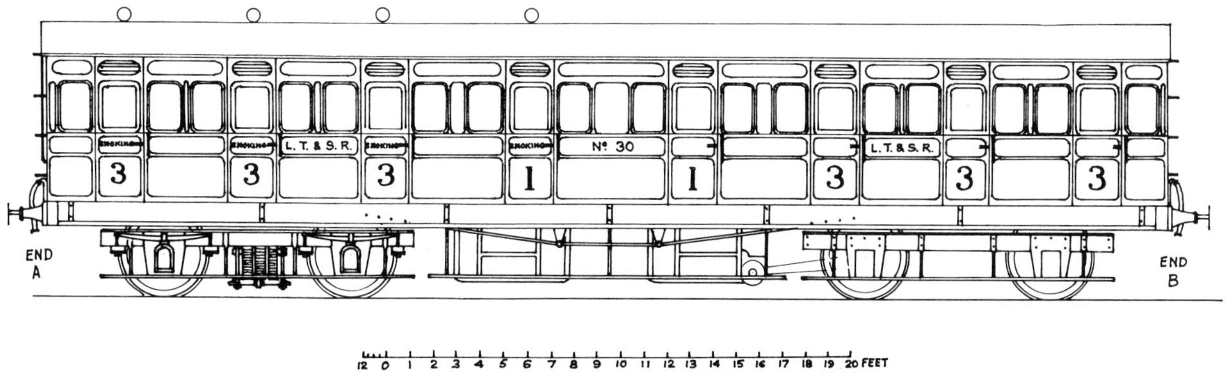

Plate 40: Composite No. 26, 48 ft in length, was also built in 1901 and 8 ft wide.

Real Photographs

For end views D and E refer to page 85.

31

DIAG. 18

THIRD CLASS.

46'0" x 8'0"

Nos. 219-234 (1901) 235-240 (1904) 258-267 (1906) 268-282 (1908) 283-294 (1910) 295-308 (1911) TOTAL 63.

FITTED WITH LAVATORIES Nos. 222. 225. 229. 233. 234. 261. 299. 306. TOTAL 8.

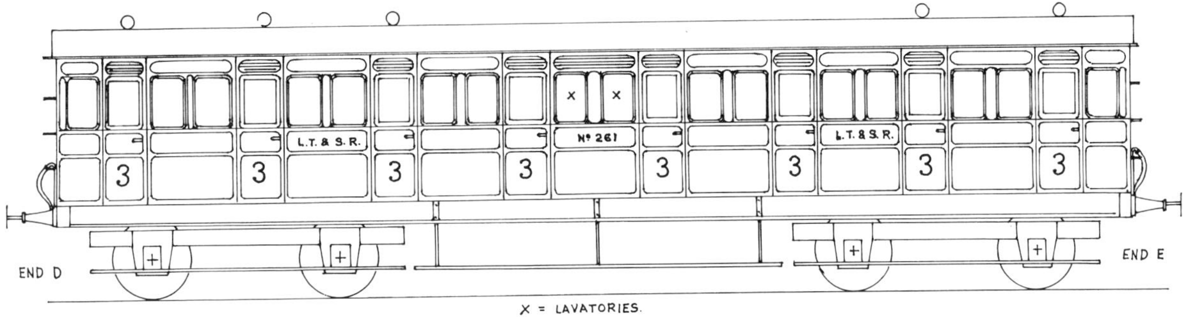

Plate 41: An eight-compartment third, No. 219, of 1901, one of the 8 ft wide carriages. *Real Photographs*

Plate 42: Standard seven-compartment first No.71; the 4th and 5th windows, from the right are frosted to accommodate the lavatories, the fifth also having a circular ventilator in the glass.
Real Photographs

32

DIAG. 6

FIRST CLASS.

DIAG. 6 WITH LAVATORIES.

Nos. 55-58 (1901) 59 (1904) 65-69, 71 (1906) 72-74 (1910) 75-78 (1911) TOTAL 18.

Nos 60 (1904) 70 (1906) WITH LAVATORIES. TOTAL 2.

48'0" x 8'4"

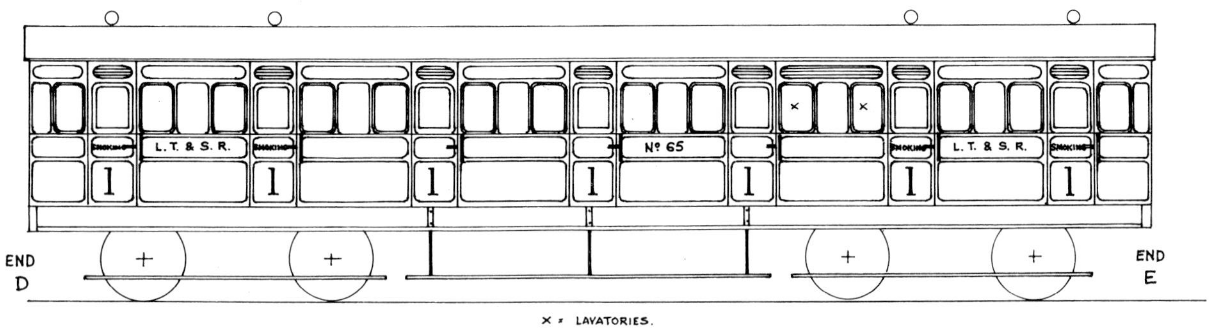

For end views D and E refer to page 85.

33

FIRST CLASS. REBUILT FROM TWO 1901 JOINT STOCK FOUR WHEELED BODIES IN 1912

Nos. 61. 62. (8 COMPTS)

DIAG. 39

52' 9½" × 8' 0"

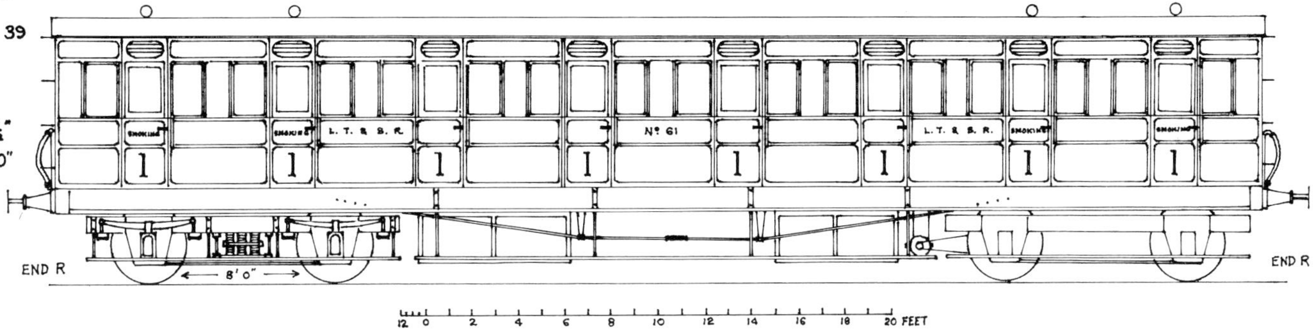

For end views R and S refer to page 61.

34

BRAKE THIRD. REBUILT FROM TWO 1901 JOINT STOCK FOUR WHEELED BODIES IN 1912. (5 COMPT THIRD + 2 COMPT BRAKE THIRD)

Nos. 59. 60. (7 COMPTS) 61-63 (6 COMPTS)

DIAG. 42

52' 9½" × 8' 0"

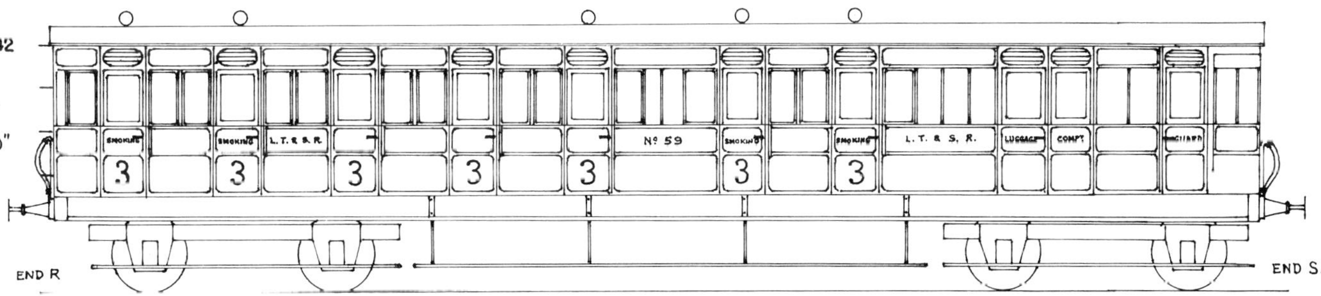

(35) THIRD CLASS. REBUILT FROM TWO 1901 JOINT STOCK FOUR WHEELED BODIES IN 1912. (5 COMPT THIRD + 4 COMPT FIRST.)

Nos. 245, 246. (9 COMPTS)

DIAG. 41

52' 9½" x 8' 0"

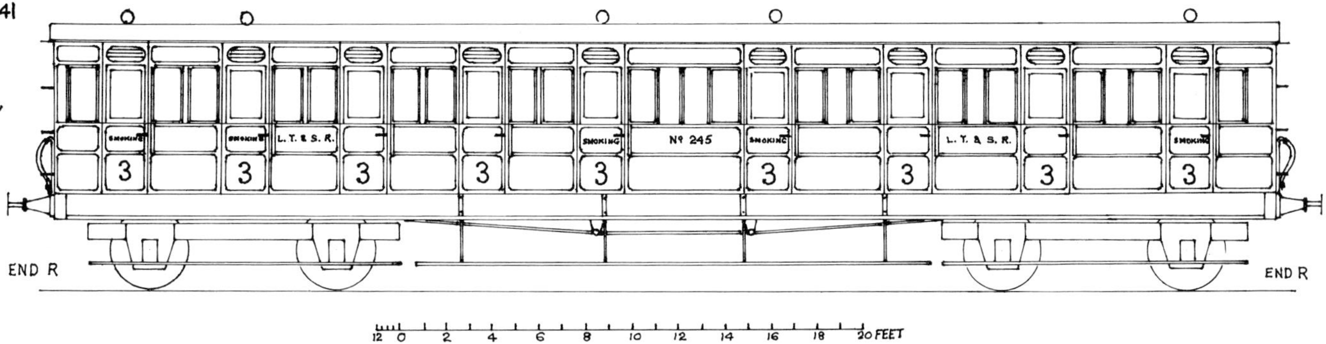

For end view R refer to page 61.

(36) THIRD CLASS. REBUILT FROM TWO 1901 JOINT STOCK FOUR WHEELED BODIES IN 1912. (5 COMPT THIRD + 5 COMPT THIRD)

Nos. 241-244 (10 COMPTS)

DIAG. 40

52' 9½" x 8' 0"

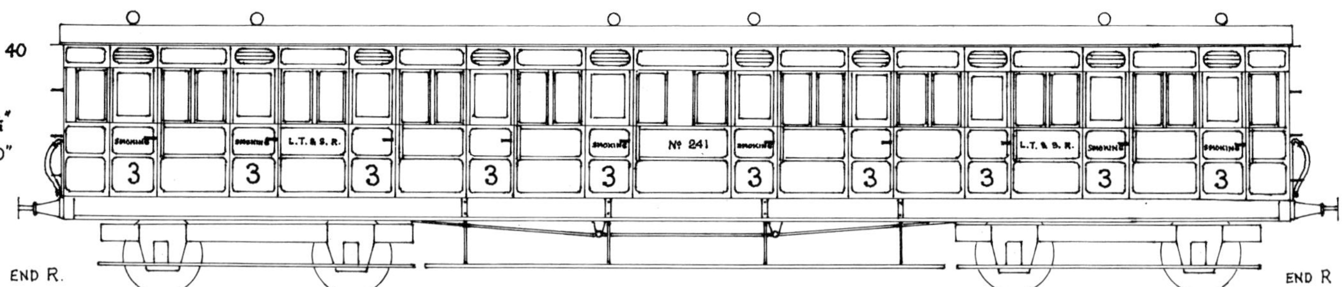

Railway between Ealing and Southend; these were delivered early in 1912. The through trains were hauled between Ealing and Barking by District Railway box-type electric locomotives (in pairs), and were handed over to Tilbury steam locomotives at the latter station. Three types of coach were used in these trains, composites, third class, and brake thirds - there were no all-first class. All were of the same general design, with sliding doors, central gangways and longitudinal seating (except against bulkheads). All were divided into two saloons by a partition one saloon being for smokers.

The two trains together comprised 10 third class, two composites, and four brake thirds; they were entirely different from any previous stock, and were designed by R.H. Whitelegg. They were 47 ft 6 in. in length and 9 ft wide, with high elliptical roofs and slightly bowed ends, standard corridor connections being fitted at each end. The access doors, at each end only, slid behind a panel decorated on the outside by a motif in coloured glass; the doorways at 2 ft 10 in. were wider than those of normal stock. The side windows were much wider than any previously used, and had in the centre of their upper portions a circular glass ventilator with five radial slots. Panelling was similar to that of the previous steam-hauled stock, but had louvre ventilators over each window. A lavatory was also included at one end. In the composite coaches a large figure '1' was painted in white in the lower part of each window in the first class saloon; smoking was allowed in the first class saloon but not in the third class. In the brake thirds the forward saloon (for smokers) was 12 ft long, and seated 16; the centre saloon (non-smoking) was 16 ft long, seating 24, and behind this was the guard's van, with a single hinged door, opening inwards, and no lookout duckets. All coaches had a small window in the ends on either side of the corridor connection, with a louvre ventilator above. Though these trains worked for a considerable distance over the District Railway, they were never jointly owned; they were always entirely LT&SR property. These corridor vehicles were generally referred to as 'the Ealing stock'.

There is some controversy over the Diagrams. One version of the Diagram list gives them as being Diagram 48 for the composites, Diagram 49 for the third class, and Diagram 50 for the brake thirds, which is logical enough. But an allegedly official Diagram which has come to light is numbered 37, and carries all three types of coach, without any differentiation in the Diagram number. To confuse matters, Diagram 37 is also given to the vacuum cleaning van No. 1856, and its offspring the loco. stores van No. 1857, which is Diagram 37A. These particular vehicles, although carrying goods list numbers, were regarded as non-passenger stock, and were given passenger list diagrams. (The non-passenger stock will be dealt with in the next chapter.) Mr R.J. Essery's *An Illustrated History of Midland Wagons* confirms these diagram numbers for the non-passenger stock, from official sources, so the position is very confused. Perhaps some day the problem will be resolved.

As far as numbering was concerned, the two composites in the Ealing stock were LT&SR No. 46 and 47, the thirds 309-318, and the brake thirds 80-83. The Midland allotted them Nos. 3126/7, 2332-2341, and 4291-4294 respectively. The LMS in 1933 changed these to Nos. 4784/5, 3066-3075, and 6397-6400, in the same order. Eight of these Ealing coaches were sold to the Government in 1940, on the cessation of the Ealing-Southend trains due to the war, and they were used for troop movements on the former Shropshire & Montgomeryshire Railway, where they remained until this War Department railway closed entirely in 1960. One or two others (or possibly part of the same batch) were noted at the Marlborough Farm (Edge Hill) army depot, working to and from Fenny Compton station.

While still on the subject of eight-wheeled coaches, mention must be made of the so-called 'rebuilds', which were constructed (probably at Derby) by mounting two four-wheeled Joint stock bodies on a new bogie underframe. Exactly how many of these were actually built is doubtful; in theory there should have been 13 in all, and this seems to be confirmed by the Midland numbers allotted, and the fact that equivalent 1933 LMS numbers were also allotted. But two numbers in the LT&SR first class list (63/4) were reserved for these coaches and apparently remained blank, only the first two (61/2) actually being filled. There is also a query about the date of these rebuilds. It is generally accepted to have taken place in 1912, and appears as such in the diagram list, but the relevant numbers in the LTS stock list which were reserved for them are all in the 1905 period. There is another theory, that when the Joint stock coaches became the LT&SR's sole property in 1905, they were renumbered into the normal stock list, after the end of the series of bogie stock built in 1904. Thus the six first class became Nos. 61-66; the 15 seconds and thirds became Nos. 241-255 in the third class list, and the six brake seconds and thirds became Nos. 59-64 in the brake third list. This all fits in very neatly with the additional bogie stock which was built in 1906, except that there are two numbers (65/6) short in the first class list, and two numbers over (256/7) in the third class list. This could be accounted for by two firsts being downrated to thirds, though when only four or five years old, it does not seem very likely. Still, when these coaches were rebuilt, in 1912 or whenever, two nine-compartment thirds, Nos. 245/6, were produced by combining a five-compartment third and a four-compartment first on the same frame. This seems to confirm the downrated-first theory, since the rest of the third class bogie coaches (241-244) were built from two, 5-compartment bodies, and came out with 10 compartments instead of nine.

Be all that as it may, the 'rebuilds' fell into five types; the first class, No. 61 and 62, were made up of two original four-compartment bodies butted together on a 52 ft 9½ in. bogie frame, the only 'surgery' required being the

For end views H and J refer to page 83.

37 EALING–SOUTHEND CORRIDOR STOCK.

THIRD CLASS. Nos. 309–318 TOTAL 10.

DIAG. 37.

47'6" x 8'6"

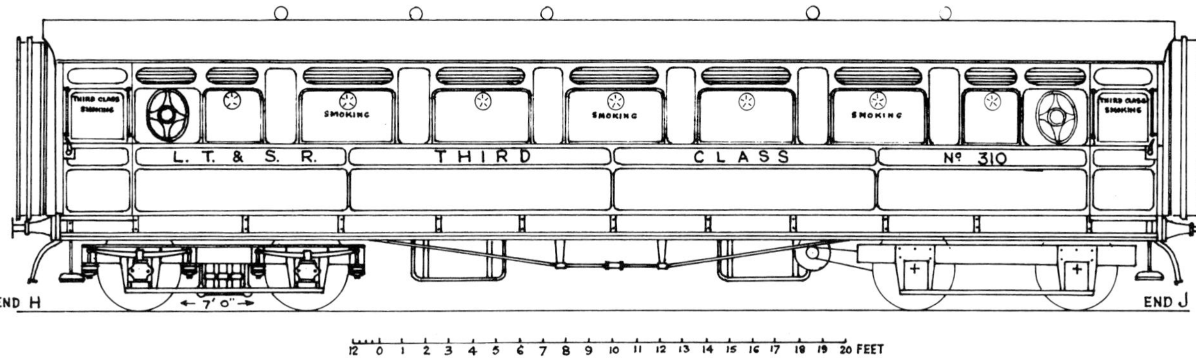

Plate 43: Brake third No. 50 (1901) had its guard's and luggage compartments in the centre.

Real Photographs

For end views H and J refer to page 83.

38

COMPOSITE.

Nos. 46.47. TOTAL 2.

DIAG. 37

47'6" x 8'6"

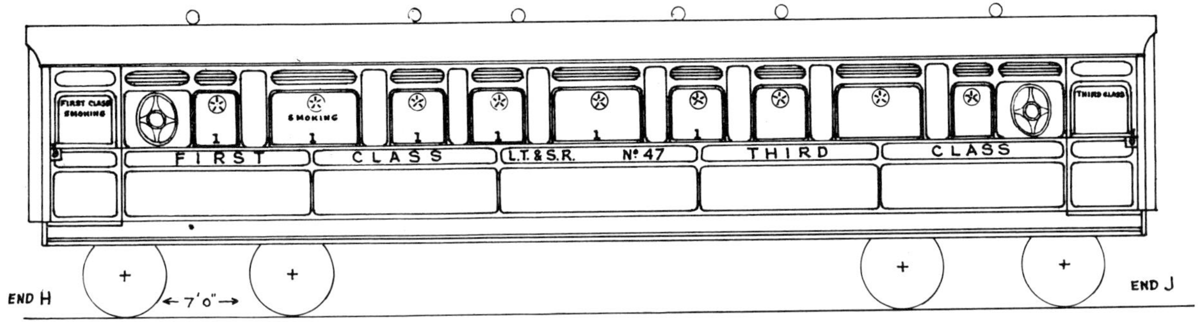

Plate 44: Composite corridor coach No. 47, typical of the two special trains for the Ealing-Southend service built in 1912.

Real Photographs

removal of the beading on the end panels which were butted together, the joint being covered by a new panel. In the main, this followed for all the rebuilds, since all were the same length, and there was no variation in the new bogie frames. The third class were also straightforward, though there were two varieties, Nos. 245 and 246 having nine compartments and the other four, Nos. 241-244, with 10 compartments, as mentioned above. However, with the brake thirds things were not quite so straightforward. There were again two varieties, Nos. 59 and 60 with seven compartments, and 61-63 with six compartments. Nos. 59 and 60 were straightforward, consisting of one five compartment third married to a two-compartment brake third. With the other three, things were not so easy, a considerable amount of chopping about being necessary to reduce the single two-compartment brake third remaining, and two of the four-compartment brake seconds, to only one compartment. This must have necessiated stripping out most of the internal partitions, and a good deal of new external panelling to produce the six-compartment brake thirds. Exactly how it was achieved is not at all clear, and it would. seem that no good illustration of one of these six-compartment vehicles is available, though photographs exist of all the other types. Hence it has not been possible to produce a line drawing. Also, for some reason, it was not given a separate Diagram, though in contrast, the two varieties of third class were on separate Diagrams.

The numbering of these coach is perhaps best tabulated:

First Class			
61, 62	Diagram 39	Midland 2607, 2608	LMS 10540/1
Third Class 10 cpt			
241-244	Diagram 40	Midland 2438-2441	LMS 14295-14298
Third Class 9 cpt			
245, 246	Diagram 41	Midland 2442, 2443	LMS 14294.
Brake Third 7 cpt			
59, 60	Diagram 42	Midland 4330, 4331	LMS 23244, 23245
Brake Third 6 cpt			
61-63	Diagram 42	Midland 4332, 4333, 4368	LMS 23241-23243

Third No. 246 was scrapped prior to 1932, and did not receive a second LMS number.

The rebuilt coaches are shown *figures 33 to 36*, which, however, should be treated with some reserve, since there are some details of dimensions which are not entirely satisfactory, and therefore may not be strictly accurate.

Plate 45: This photograph taken in March 1956, shows the livery of a restored coach in LT&SR colours. *H.C. Casserley*

Plate 46: Axle box and wheel details of LT&SR 4-wheeled third class coach built in 1876. Seen here on the Corringham Light Railway on 17th May, 1947. *H.C. Casserley*

Directors' Saloon

A sumptuous saloon was built at Plaistow in 1912 for the use of the LT&SR Directors, but as a result of the Midland take-over, it was hardly ever used for its specific purpose, and was transferred to Derby for departmental use. It was a large vehicle, 46 ft 9 in. long and 9 ft 6 in. wide, with a deep elliptical roof having domed ends. Both sides and ends had a pronounced tumblehome, and the ends themselves were bowed, with two large windows. Most of the side windows were wide, with curved top edges; both sides were not alike, as can be seen from the drawings (*figures 40, 41*). Waist panels had semi-circular ends, and the lower panels had rounded upper corners and square lower ones. Most of the cant rails were taken up by long louvre ventilators.

The dining saloon was 20 ft 8 in. long and seated 10 persons at normal dining tables. However, the lounge next door (at one end of the coach) could easily be fitted with dining tables, and would provide accommodation for a further six. These two saloons were finished in Spanish mahogany, and the floor was covered by Axminster carpet on a layer of felt. More or less in the centre of the coach was a kitchen, approximately six feet square, containing an electric water heater, electric kettle, grill-hotplate, the main switchboard, and an electric fan. There were also cupboards, an ice box, and a sink. How all this equipment was fitted into a 6 ft square space, and still leave enough room for the attendants to work in, is a mystery. The lavatory, next to the kitchen, was lined with Italian marble, the wash basin and other equipment was nickel plated, hot and cold water being provided. The kitchen, lavatory, and corridor floors were covered with Silvertown rubber tiles.

At the far end of the coach was a smoking room, 12 ft 3 in. long, furnished with seven basket-work chairs covered in brown leather. This saloon was panelled in dark oak, with inlaid Sheraton work. The entrance vestibule, which extended across the coach, was fitted with wardrobes and an umbrella stand. Against both ends of the vehicle, in the centre, was a fireplace, with mantlepiece and polished wood surround, fitted with an electric fire. The electric power consumption of this coach in full use was very considerable and was provided by two axle-driven dynamos of 3½ kw rating with a bank of 30 Chloride storage cells. Both Vacuum and Westinghouse brakes were fitted, also steam heating .

The underframes and bogies were painted royal blue, with the bodywork in the standard varnished teak, with white roof. Lettering, however, was reduced to 'M.R.' in script form on the waist panel on each side. The only suggestion of a number was on the building plate on the solebar, and this cannot be established with any certainty. It is reputed to have been numbered 4 in the saloon list; this cannot be definitely established either. All that

For end views H and J refer to page 83.

39

BRAKE THIRD. DIAG. 22 Nos. 80-83 TOTAL 4.

DIAG. 37.

47'6" x 8'6"

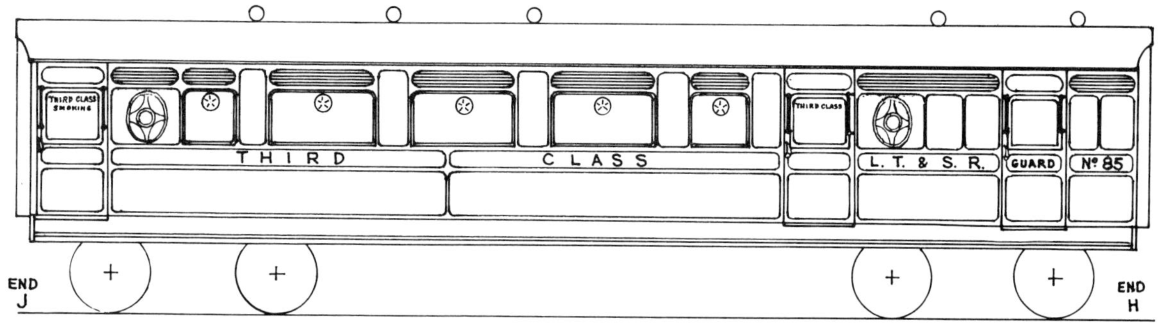

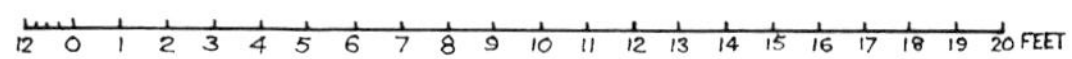

40

DIRECTORS' SALOON.

DIAG. 43

46'9" x 9'0"

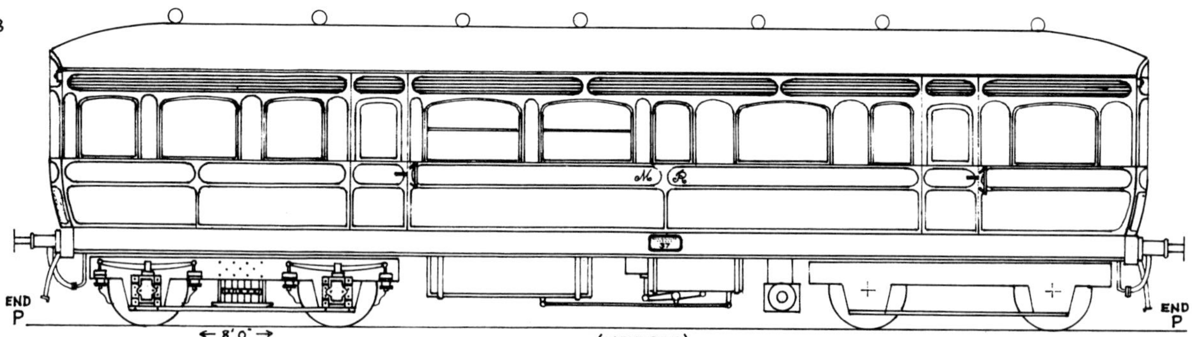

(NEAR SIDE)

For end view P refer to page 85.

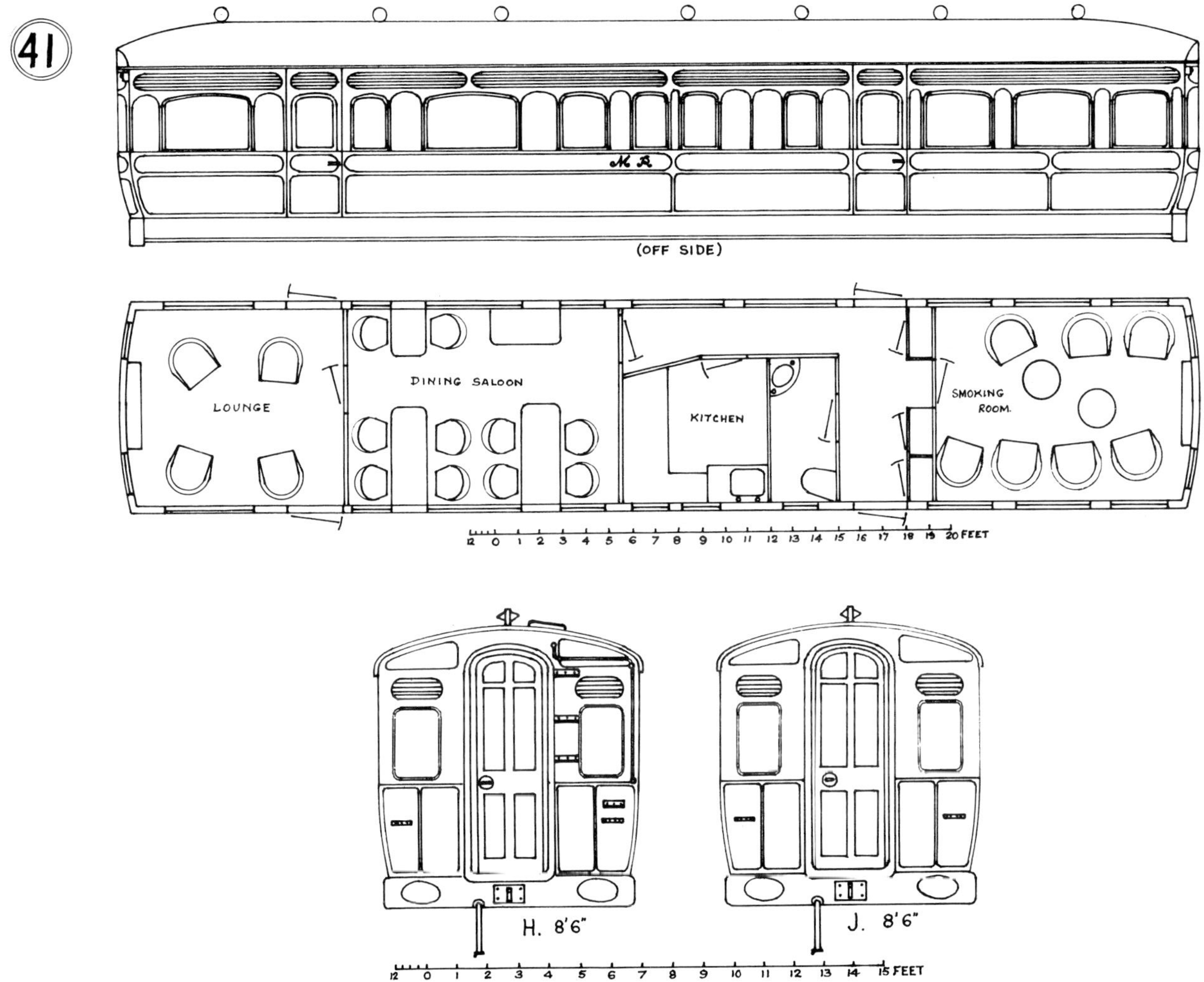
41
M R
(OFF SIDE)
LOUNGE
DINING SALOON
KITCHEN
SMOKING ROOM.
12 0 1 2 3 4 5 6 7 8 9 10 11 12 13 14 15 16 17 18 19 20 FEET
H. 8'6"
J. 8'6"
12 0 1 2 3 4 5 6 7 8 9 10 11 12 13 14 15 FEET

is definitely known is that it was MR 2799, and in 1933 received the LMS number 817, in the series of public-use saloons, so in all probability it had undergone considerable internal alterations by that time.

After the electrification of the District Railway in 1905, a proportion of that company's electric stock appeared in the books as being owned by the LT&SR; this continued as further electrification took place, and on the extension to Upminster in 1932, the number of such coaches was given as 110. They were all standard District Railway design, and were numbered in their series; they never had any LT&SR numbers, and carried no other identification. Nevertheless, they were given an LT&SR Diagram, - 44.

Plate 47: The first trains employed full brakes like this one, numbered 5; only 12 were built before third-brakes became the norm. *Real Photographs*

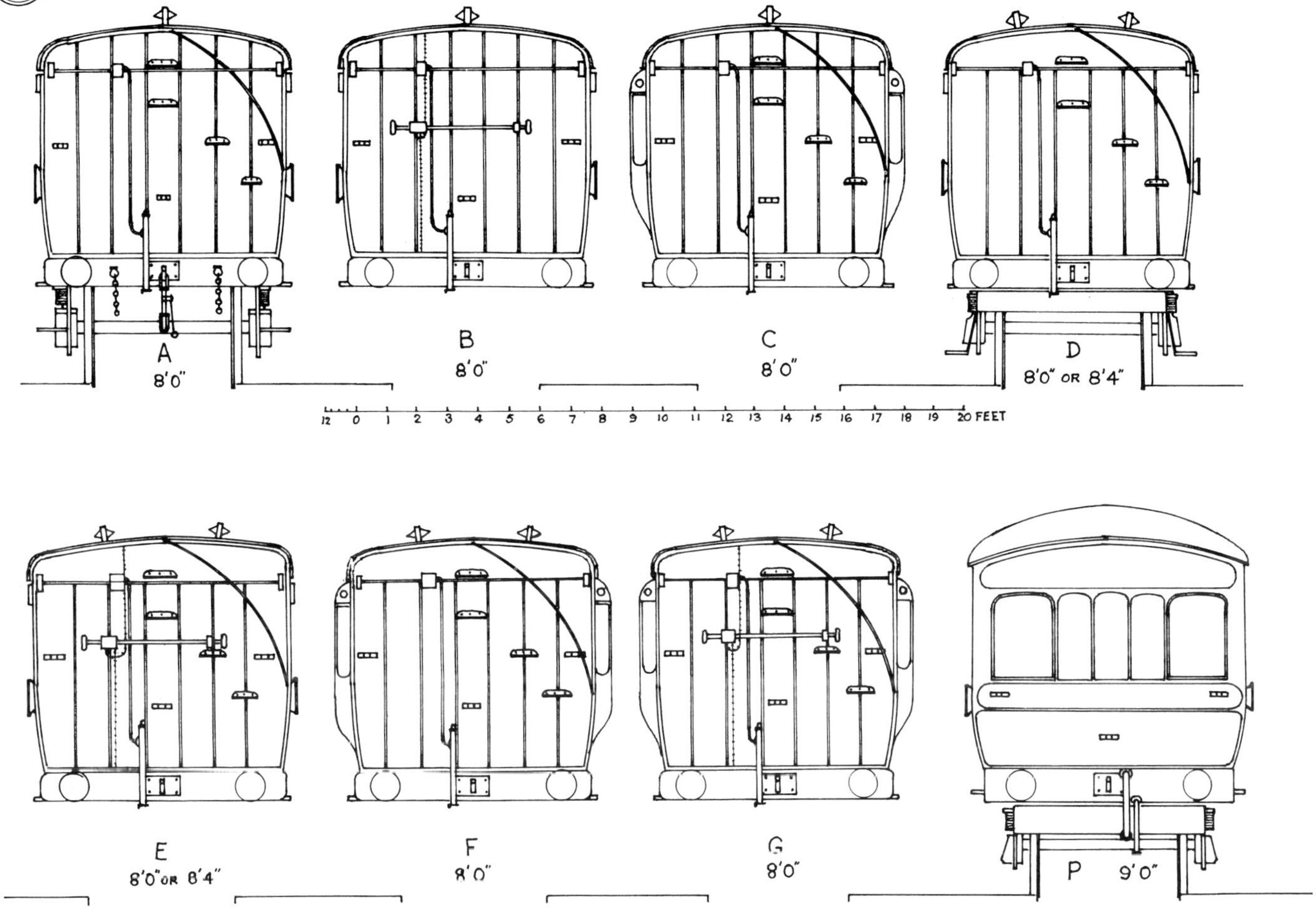
A
8'0"
B
8'0"
C
8'0"
D
8'0" OR 8'4"
12 0 1 2 3 4 5 6 7 8 9 10 11 12 13 14 15 16 17 18 19 20 FEET
E
8'0" OR 8'4"
F
8'0"
G
8'0"
P 9'0"

Plate 48: A short horse box (16 ft) of Diagram 31; No. 3 of the 1878 batch, with large groom's compartment. *Real Photographs*

Plate 49: Diagram 32, the longer version of horse box (17 ft 6 in.) No. 7 of the 1902 batch with short groom's compartment and locker at other end. Note the unusual plate axle-guards, instead of the usual 'W' irons. *Real Photographs*

Chapter Four

Non Passenger Carrying Stock

Almost every railway company had a number of different vehicles intended for various purposes, which did not carry passengers, but were designed to run with passenger trains, and were consequently fitted with automatic brakes and painted in passenger livery. The LT&SR was no exception, and had a considerable number of vehicles which fell into this category, mostly numbered in separate lists according to type.

Passenger Brake Vans

The earliest trains had separate vehicles for the accommodation of the guards, and for carrying parcels and other merchandise. These vans had a projecting lookout on each side (known as duckets) to enable the guard to see along the train, and to observe signals. The guard was also provided with a hand brake wheel somewhere in the van, by which he could apply the brake in an emergency. In the course of time it was thought more economical to make the brake van carry a few passengers, and so two or more ordinary compartments were added to the van; thus the brake third was born. These became the norm on the LT&SR, and no more full brake vans were obtained. There was only one type of full brake, with four wheels, 26 ft long and 8 ft wide (9 ft over duckets) to Diagram 29 (*figure 15*). The duckets were in the middle of the van, with the guard's door adjoining. In each half of the van were a pair of double doors for convenience in loading, large objects. Apart from normal drop lights in the doors, there were no windows. Seventeen of these brake vans were built, Nos. 1-7 in 1878, Nos. 8-10 in 1882, and the final seven, Nos. 11-17, in 1886. They were all renumbered into the Midland list in 1912, as Nos. 165-168, 170, 172, 174-178, 187, 191, 193, 194, 202 and 206.

Horse Boxes

The conveyance of horses was always a considerable part of the railway's operations, particularly before the days of the motor vehicle, and it was not until the early 1920s, when motor traffic came into its own, that the use of railway horse boxes began to decline. Being only a short railway, the LT&SR had only 12 horse boxes, though on some of the larger railways the numbers ran into hundreds. The Tilbury stock was covered by two diagrams, of which the smaller version (Diagram 31) were 16 ft in length and 7 ft 9 in. in width, running on four wheels. Nos. 1-6 were built in 1878; at one end was a groom's compartment, six feet long, with a door in the centre and window

DIAG. 31 HORSE BOX.

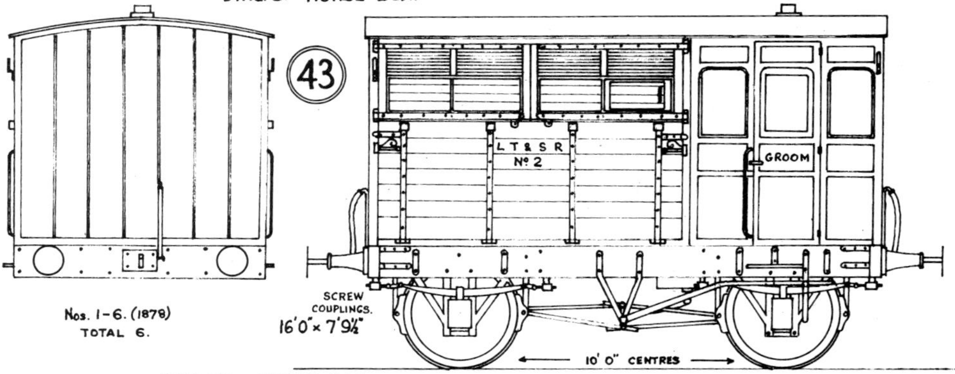

Nos. 1–6. (1878)
TOTAL 6.

16' 0" × 7' 9½"

DIAG. 32. HORSE BOX.

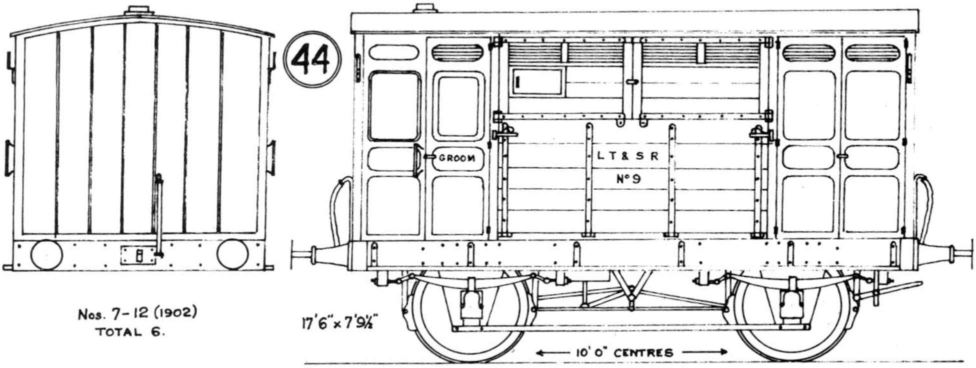

Nos. 7–12 (1902)
TOTAL 6.

17' 6" × 7' 9½"

DIAG. 34 PASSENGER LUGGAGE VAN.

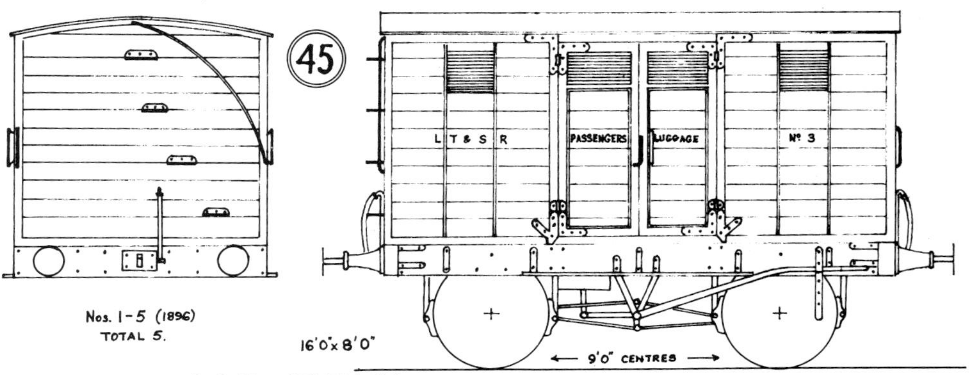

Nos. 1–5 (1896)
TOTAL 5.

16' 0" × 8' 0"

DIAG. 35. BULLION VAN.

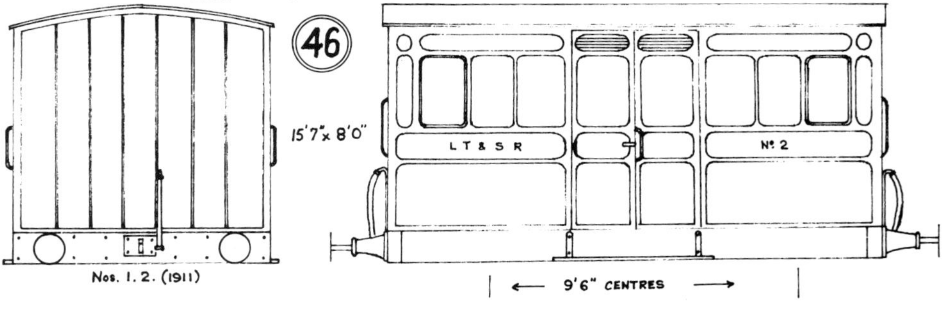

15' 7" × 8' 0"

Nos. 1. 2. (1911)

each side of it - more or less a normal passenger compartment. The horse section occupied the rest of the vehicle, with three stalls arranged lengthwise. There were two doors in the upper part of each side, with a large drop-down door below, (which served as a loading ramp). The underframe was the normal wagon type with 'W' irons and short springs, the wheelbase being 10 feet. The Westinghouse brake and screw couplings were not fitted at first; these were added some eight years later when the air brake was made standard. Only one oil lamp was fitted, in the groom's compartment; when the change to electric lighting was made, one was also fitted in the horse section.

A slightly larger horse box came out in 1902; these were Diagram 32, and were 17 ft 6 in. long, though the width remained the same, 7 ft 9½ in Nos. 7-12 were somewhat differently arranged, having a shorter groom's compartment, only 4 ft long, with door and only one window. The horse section was as before, but at the other end of the vehicle was a storage locker, 4 ft long, for fodder and harness equipment. Outer appearance and fittings were similar to the earlier design, but these had full length footboards at the bottom of the solebars, and plate axleguards instead of 'W' irons. Wheelbase was 10 ft, as before, and Westinghouse brakes and screw couplings were fitted when new. The two types of horse boxes are shown in *figure 43 and 44.*

Match Wagon

This was a peculiar special list vehicle which was designed as a means of enabling District Railway electric stock to be hauled by a steam engine if need arose, through failure of the current supply or other breakdown. Only one was built, at Plaistow, and it was rarely called upon for use. Essentially it was an ordinary flat wagon, with deeper than normal solebars; it was fitted with Westinghouse brakes, and had at one end an ordinary set of buffers and screw coupling. At the other end it had a special automatic coupling matching those of the District Railway, and a pair of short buffers. Thus it could be coupled between a steam locomotive and the electric stock. It was 16 ft long and 8 ft wide, and was built in 1907 with the number 1812. There was a rather ambiguous treatment of non-passenger stock on the Tilbury line; mostly they were numbered in separate series according to type, but here and there for no apparent reason, some non-passenger vehicles were numbered in the goods list, though appearing on passenger diagrams. The match wagon was a case in point. It was on Passenger Diagram 30, and is shown in *figure 48.*

CARRIAGE TRUCK DIAG. 33 15'6"x8'0" Nos. 1-6 (1878)
TOTAL 6.

47

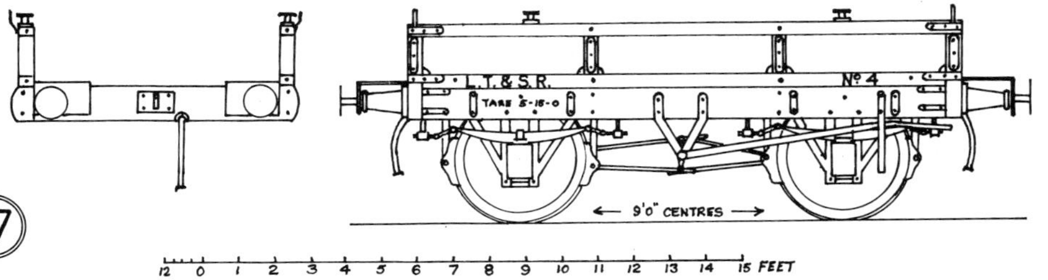

Plate 51: Open carriage truck no. 6, of 1878 (Diagram 33). Of conventional design, but evidently not a great deal called for, since only the one batch of six was ever built.
Real Photographs

48

MATCH WAGON. DIAG. 30.
16'0" x 8'0"

No. 1812 (1907)
ONE ONLY.

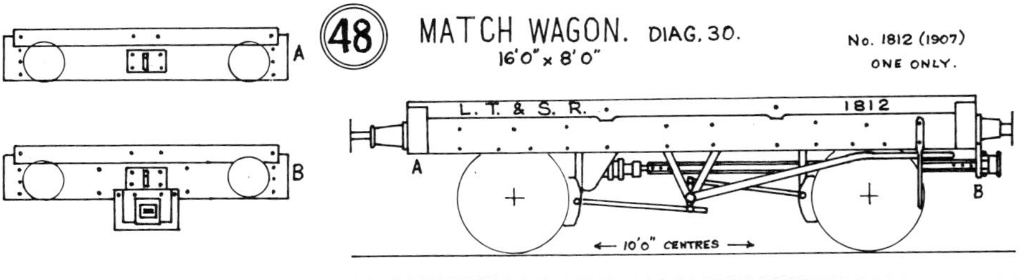

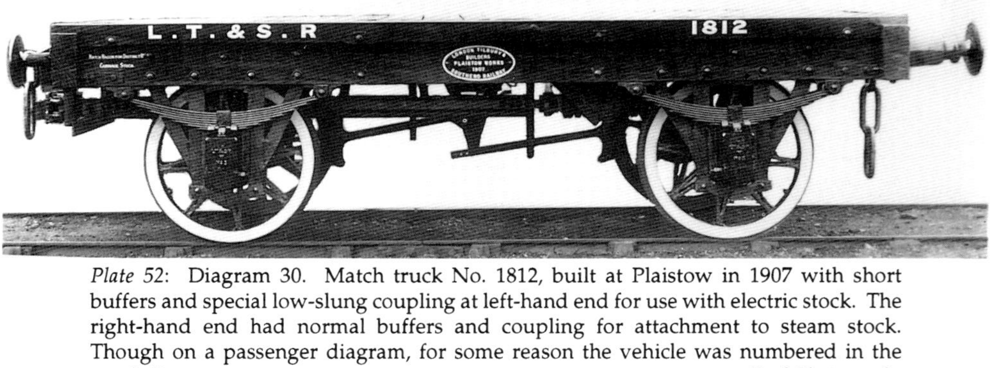

Plate 52: Diagram 30. Match truck No. 1812, built at Plaistow in 1907 with short buffers and special low-slung coupling at left-hand end for use with electric stock. The right-hand end had normal buffers and coupling for attachment to steam stock. Though on a passenger diagram, for some reason the vehicle was numbered in the goods list.
Real Photographs

Open Carriage Trucks

Once again there were six of these, numbered 1-6 and built in 1878 by the Metropolitan C&W Co. Apparently there was not much call for them, since no more were built (Diagram 33, *figure 47*). They were of the usual type, with one bar on each side, 1 ft 6 in. above the solebar, and supported on four stanchions. They were 15 ft 6 in. in length, and 8 ft wide. They had normal underframes with one-side hand brakes, but Westinghouse brakes were also fitted.

Diagram 34 Passenger Luggage Van (*figure 45*)

These were somewhat unusual vans, of a totally different design to the ordinary goods vans, but which they strongly resembled, though the details were entirely different. They were 16 ft long and 8 ft wide, with 11 planks each side. In the centre was a pair of double hinged doors, 5 ft wide, which had louvres in the upper part. The remainder of the sides had a louvre ventilator in the centre of each section, with a vertical narrow moulding each side of the ventilator. The iron strapping was of quite unorthodox design. For some reason these vans had four footsteps and a curved handrail at one end; as there were no lights or other fittings on the roof, the provision of roof access seems pointless. The underframe was of standard wagon pattern, with one side hand brake, and Westinghouse equipment. General dimensions were length 16 ft 0 in. width 8 ft 0 in., wheelbase 9 ft 0 in. Five of these vans were built by Brown, Marshall & Co. in 1896, numbered 1-5.

Diagram 35 Bullion Van (*figure 46*)

Only two of these special vans were built, numbered 1 and 2, at Plaistow in 1911. They were only 15 ft 7 in. long, and 8 ft 0 in. wide, with a wheelbase of 9 ft 6 in. The bodywork was the same as the ordinary passenger coaches, with rounded corners to the panelling. A window was fitted at each corner, but the doors had solid panels. There was an entrance vestibule which went right across the van, and each end section was virtually a large safe, lined with armoured steel, and fitted with a safe door (with special locks) in the middle of the vestibule on either side. Again, these vans were not often used, but came into service on odd occasions when gold bars or other valuable material had to be transported to or from ships in Tilbury docks. There were no lights, and no access to the roof. Westinghouse brakes were fitted, since these vans were always attached to passenger trains. The underframe was the standard wagon pattern, with short springs and 'W' irons.

Plate 53: Accumulator van, Diagram 36. Only one of these was ever built, in 1905. Designed for conveying batteries for passenger stock to and from Plaistow Works. *Real Photographs*

Diagram 36 Accumulator Van (*plate 53*)

A single example of this type was built at Plaistow in 1905, for transporting carriage batteries. In general construction it was a standard goods van, with 12-plank sides, but with a pair of carriage doors in the centre of each side. It had two torpedo ventilators on the roof, and louvre ventilators in the top of each door, plus a louvre ventilator in each end, since batteries could give off noxious fumes under certain circumstances. The doors also had normal drop lights, and there were footsteps and handrails at the ends. It was 16 ft long and 8 ft wide, numbered 1, and assigned to the Locomotive Department. An either-side handbrake was fitted, of an unusual type, with short levers operated from the ends of the vehicle.

Diagram 37 Vacuum Cleaner Van (*figure 49*)

One of this type of van was built at Plaistow in 1912, numbered in the goods list as 1855. It was almost entirely of metal construction, with a dropped floor between the axles; all external strapping was of 'T' section. It was 25 ft long and 8 ft wide, with a wheelbase of 16 ft. Not much daylight penetrated the inside of this van, since there were only two small square windows in each side, located over the axles; the sliding door was of sheet metal on a channel section frame. Five metal louvres on each side of the windows provided the only ventilation. Inside was an electric motor driving a powerful suction fan which deposited the rubbish into removeable bags for subsequent disposal. Outside the fan was connected to a long flexible pipe which could be taken wherever need arose. This van was taken eventually to all sorts of odd places on the LMS system after the Grouping for various extensive cleaning jobs - coaches, stations, sheds and so forth. (It was noted at Windermere in 1966.) For some reason, the van was apparently renumbered 1857 about 1916.

Diagram 37A Loco. Stores Van (*figure 50*)

Some mystery surrounds this van; it came out of Plaistow works in 1916, but whether built new, or a conversion of a second vacuum cleaner van is not clear. It was obviously constructed to the same drawings as the cleaning van, but the middle section was open, and the windows were omitted. The two end sections had a door in the middle of the partition, which could be padlocked. The open section in the middle was fitted with two shallow drop down doors on each side to facilitate loading bulky articles. It was numbered (or renumbered) 1856, was allotted to the Loco. Department, and was used for the transportation of various stores and spare parts, mainly within the

49

VACUUM CLEANING VAN. DIAG. 37

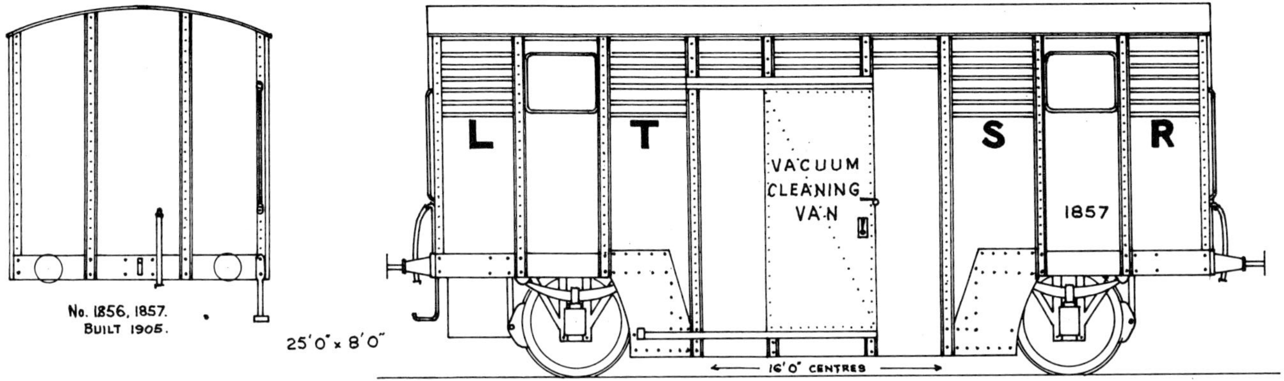

No. 1856, 1857.
BUILT 1905.

25'0" x 8'0"

50

LOCO. STORES VAN. DIAG. 37A.

REBUILT FROM VACUUM CLEANING VAN 1916.

No. 1856 ONLY.

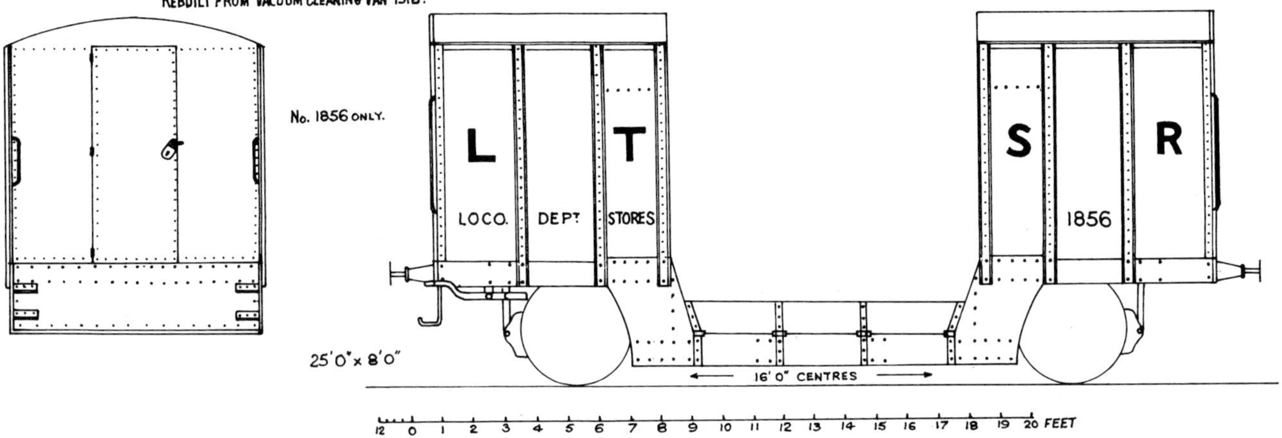

25'0" x 8'0"

Plate 54: Diagram 37. The unique vacuum cleaner vans, of all metal construction. Numbered in the goods stock list, although allocated a passenger diagram.

Real Photographs

Plate 55: Loco. Stores wagon, Diagram 37A. Built (or rebuilt?) from the same drawings as the vacuum cleaner vans. The origin of the vehicle is in doubt; it could have been built new, or it might have been rebuilt from vacuum cleaner van No. 1856.

Real Photographs

Plaistow Works/shed area. It was not Westinghouse-fitted - the vacuum cleaner van was. At the same time (1916) the crane runner took the number 1855, it having previously carried a Loco. Dept. number.

Diagram 38 Theatrical Scenery Van (*figure 51*)

Built by Cravens Ltd. in 1911, these eight vans where six-wheeled, and were intended for transporting theatrical scenery flats and other properties. They were given numbers 6-13 in the same series as the passenger luggage vans of Diagram 34. The bodies were built of wood with vertical planking on inside frames. Two wide sliding doors were fitted on each side, with three windows, one in each section, the glass having horizontal brass bars on the inside. Each end had a pair of large hinged doors, and iron plates over the buffers to allow vehicles to be run into the van - they were also used occasionally as covered carriage trucks. Three lamps were provided in the roof, and Westinghouse brakes were fitted in addition to the usual hand brake. These vans were 34 ft long and 8 ft wide, and had a wheelbase of 21 ft, equally divided.

The Midland Railway numbers allotted to the above vehicles are not known, since unlike most other railways, who simply added so-many thousands to the original numbers, the Midland pursued a policy of numbering acquired vehicles into any convenient gaps in the main list, so that there was no continuity. From the historian's point of view, this was a very bad policy, since there were no guide lines whatever to the new numbers, and in the absence of any lists, tracking down the individual numbers was a matter of sheer luck. Unfortunately, this policy also applied to the goods stock.

51

SCENERY VAN. DIAG. 38.

34'0" × 8'0".   Nos. 6-13 (1911)   TOTAL 8.

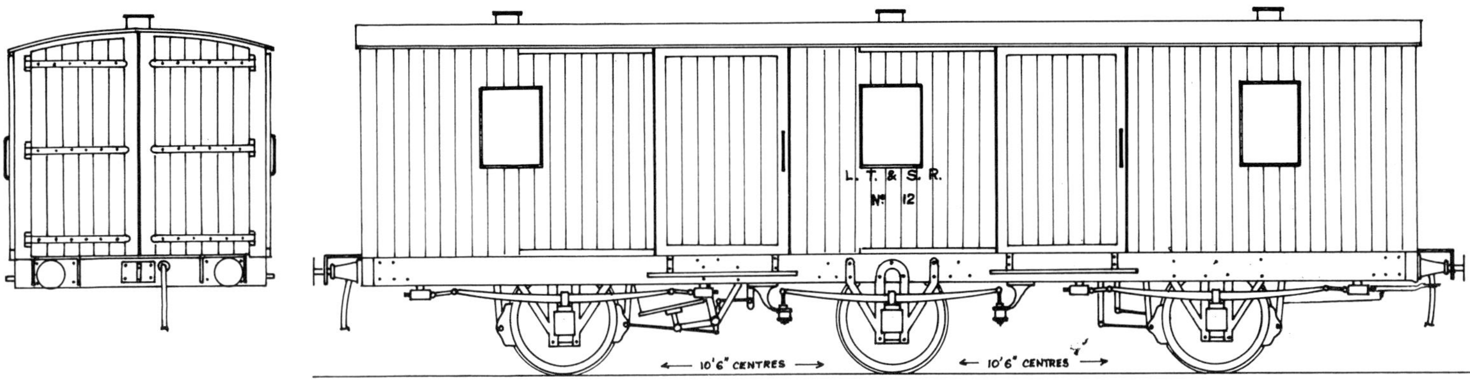

Plate 56: Theatrical scenery van No. 11, one of a batch of eight built in 1911. It had sliding doors in each side, and large double doors in each end. Used also as covered carriage trucks (Diagram 38).
Real Photographs

Chapter Five

Goods Stock

Goods rolling stock, unfortunately, has been sadly neglected by railway historians, often being dismissed - if mentioned at all - in a short paragraph. As a result, the information available generally, has been very scanty. However, with the LT&SR we are on much firmer ground, for by sheer luck one document pertaining to the goods stock of the company came to light during Mr R.J. Essery's researches into the rolling stock of the Midland Railway; thus his two-volume history of Midland wagons has been of great advantage to me in producing this book, and I have to record my most grateful thanks to Mr Essery.

Like the passenger stock, a series of Diagrams was issued to cover the goods vehicles, and these have a very fair authenticity, unlike the passenger Diagrams. The Diagram numbers apparently ran from 1 to 19, but there are four blanks in this series (5, 7, 8, 14) which may or may not have been used at all, or may have lapsed through vehicles of the particular type having all been withdrawn. There is nothing at all available to support or disprove either of these hypotheses, so it has to be left up in the air. Perhaps some day something may be unearthed to point us in the right direction, but it seems highly unlikely.

The Diagrams do not appear to have been issued in chronological order; probably they were allotted when it was first decided to issue a Diagram Book, which would certainly not have been in 1880, it is more likely to have been about 1900 - or possibly even later. For instance, Diagram 19 covered the fire-engine truck, which dated from 1883, while Diagrams 17 and 18 dated from 1904. There is, however, the possibility that Diagram l9 was not included at all, owing to its particularly special nature (most of the breakdown train vehicles were not in the Diagram Book either.) Perhaps Diagram 19 was allotted very late on, a theory confirmed by the running number given to the vehicle (1853) which dates from 1907. The two breakdown cranes and the other vehicles which comprised the breakdown train did not appear in the Diagrams, and it seems rather arbitrary that only two of them were given numbers in the goods stock list (1854 and 1855, dating from late 1907) while the rest were numbered as 'Loco. Dept. No. 1 , 2, 3' etc.

When goods wagons were taken out of stock, either as worn out or after damage in an accident, they were usually replaced almost immediately by new wagons of the same type and bearing the same numbers. Such replacements were built at Plaistow, and if nothing else, served to keep the carriage & wagon works ticking over. So for example, in any particular series of consecutive numbers, the building date of any of them could only be found by reference to the number plate on the solebar; the fact that the particular series was built in - say - 1885 no longer held good. Some wagons were rebuilt to a different design, for example, some ballast wagons became open goods wagons; 15 goods vans were rebuilt as meat vans, as were also two cattle vans; two cattle vans became fish vans, and perhaps the most astonishing of all, 12 goods brake vans were rebuilt as open wagons. In most cases the rebuilds retained their original numbers, but in the case of the rebuilt brake vans, these were renumbered, since goods brakes were numbered in a separate series.

Most LT&SR goods wagons had brakes on one side only, some of the older ones indeed having only one wheel braked. The brake levers were of a variety of shapes - even within the same series - straight, curved, long, short, in any sort of combination. A small proportion of the stock had brakes on both sides. The livery was similar to that of the Midland Railway, lead grey with black ironwork. Lettering was in a pale yellow (perhaps best described as 'off-white') with black shading. There were some variations from this, however, gunpowder vans were bright red with white lettering, and fish vans a pale colour which was difficult to describe; it has been called pale blue, or alternatively pale green. These had black lettering. Some timber trucks and odd vans also had black lettering, for no apparent reason. Rather unusually, goods brake vans were classified as departmental stock.

Very, very few of the Midland numbers allotted to the LT&SR stock in 1912 are known. With the exception of the 25 gunpowder vans, of which the Midland numbers are known in their entirety, only odd ones here and there have come to light. This, as has been mentioned before, comes from the Midland policy of numbering acquired vehicles into any convenient gaps which were available.

The goods stock will be dealt with in Diagram order.

Diagram 1 Open Wagon (*figure 52*)

These all had five plank bodies with a central drop down door five feet wide in each side. They were rated to carry 10 tons - in fact, there were very few Tilbury wagons which had a different rating. The wheels were supported in standard 'W' irons, and were spaced at 9 ft centres. Length over body was 16 ft, and width 8 ft. In all 800 were built over a period of 28 years, by four different makers, while a further 12 were rebuilt at Plaistow from goods brake vans. The total was further increased by the modification of 40 Diagram 12 ballast wagons to open goods wagons in 1898. There were various minor differences between the batches, such as the metal strapping and the position of the bolts in the corner plates, for example. There were also variations in the lettering and numbering; some had the number painted on the bottom plank in the right hand corner, while others had the number on the top portion of the solebar. Most of the wagons had the standard off-white lettering, but a few had it in black. Brake gear varied in arrangement, some having long handles and others short ones. Some batches had iron stop plates to prevent the doors dropping too far, but others did not. As with most other types, a number of the older wagons were replaced from time to time by new ones built at Plaistow bearing the same numbers. The full list is as follows:-

Nos. 101	-	200	built	1876	by	Lancaster C&W Co.
201	-	250		1879		Metropolitan C&W Co.
583	-	682		1888		Midland C&W Co.
683	-	782		1891		Midland C&W Co.
887	-	898		1898		Plaistow (converted from Goods Brakes)
1044	-	1143		1898		Metropolitan C&W Co.
1150	-	1199		1899		Metropolitan C&W Co.
1300	-	1399		1900		J.A. Abbott & Co.
1400	-	1499		1900		Metropolitan C&W Co.
1675	-	1774		1904		Metropolitan C&W Co.
251	-	290		1898		Plaistow (converted from Ballast wagons)
						Total 852

Diagram 2 Goods Brake Van (*figure 53*)

The design of these was unusual. They had an open verandah at each end, but without corner posts, so that the roof was unsupported at its outer ends. The sides had 11 horizontal planks, with framing outside, including a diagonal strut from the centre of the roof to the bottom corner at each end. The end panels were also very solid, with external framing and diagonal struts similar to the sides. At each end of the van portion there was a door in the centre of the bulkhead, with a window in the upper half, the windows in the doors being the only source of light, except for an oil lamp in the middle. Later some of these vans had a window cut in the bulkheads on either side of the door. No chimney was shown on the diagram, so whether the vans were fitted with stoves is a debatable point. No chimney is visible either on any of the few available photographs. A footstep ran the full length of the vehicle at axle level, with a short upper footstep under each verandah at the bottom of the solebar. A handrail was fitted, running the full length of the van, about a foot above the solebar, and joining the lower part of the vertical handrails at the verandah. Brakes operated on all four wheels, worked by a wheel and screw inside the van. These vans were rated at 10 tons. They varied slightly in length from 16 ft to 16 ft 3 in., and were 7 ft 11 in. wide over framing. They had the standard wheelbase of 9 ft. The number was carried

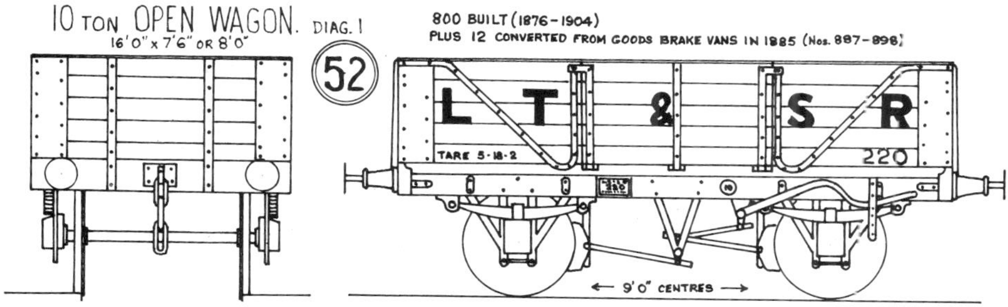

Plate 57: Standard goods brake van No. 34, Diagram No. 2, built in 1884. Of somewhat unusual design, with unsupported roof ends, the stock total of these vans had reached 46 by 1885, and was maintained at this level in spite of several being rebuilt for other uses. New vans were built at Plaistow to replace those lost. *Real Photographs*

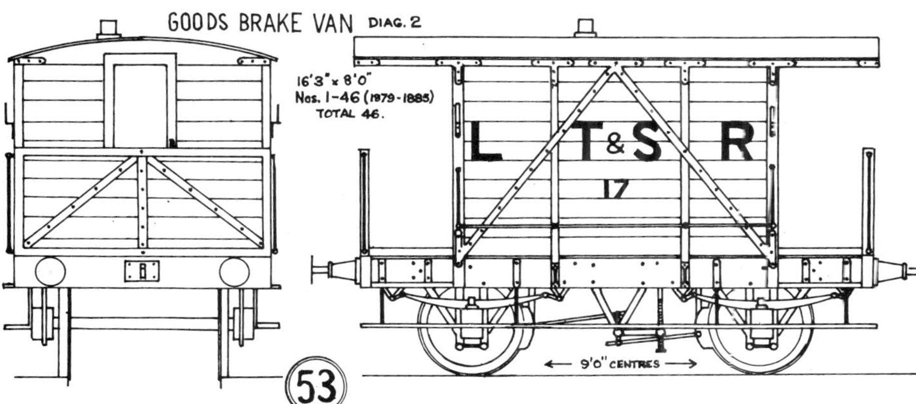

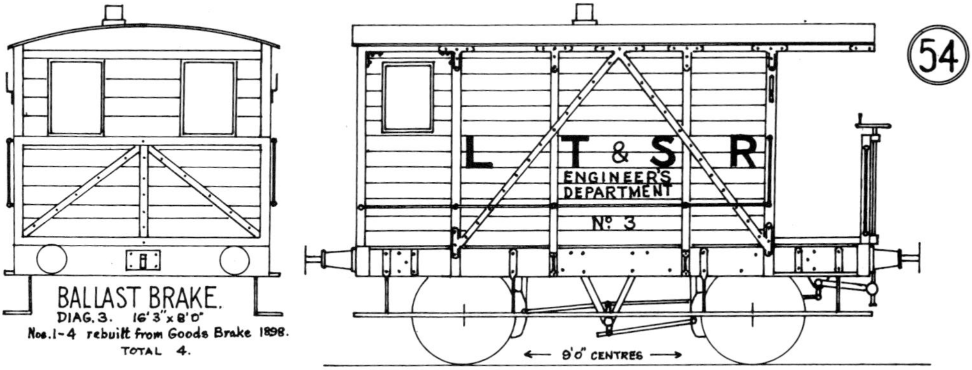

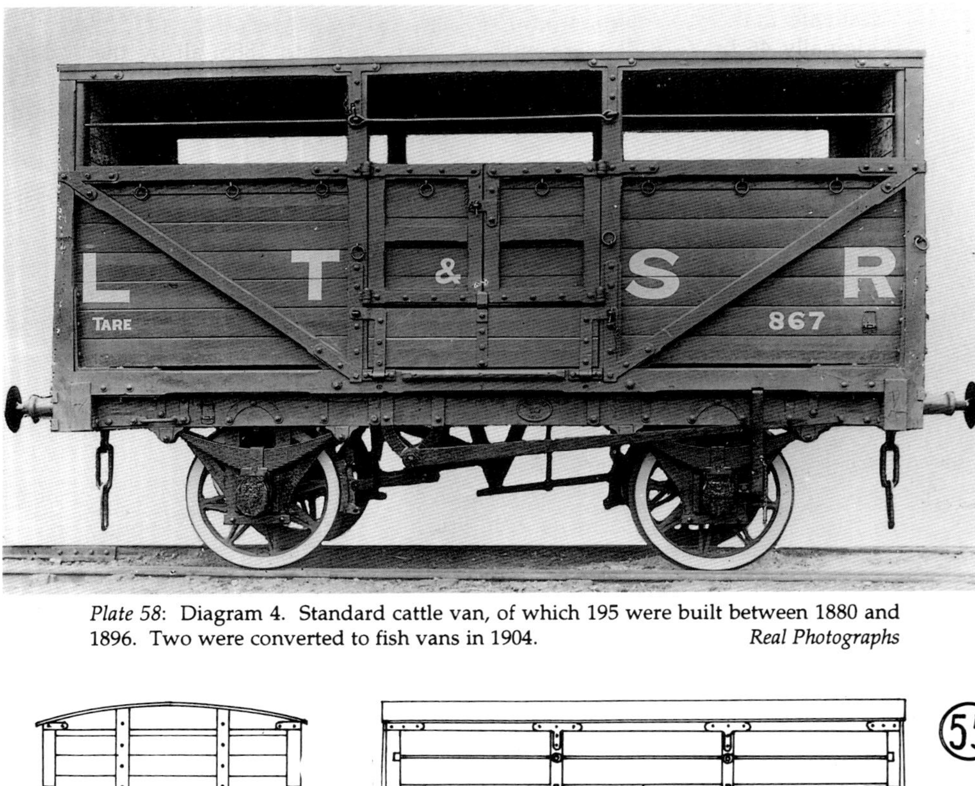

Plate 58: Diagram 4. Standard cattle van, of which 195 were built between 1880 and 1896. Two were converted to fish vans in 1904. *Real Photographs*

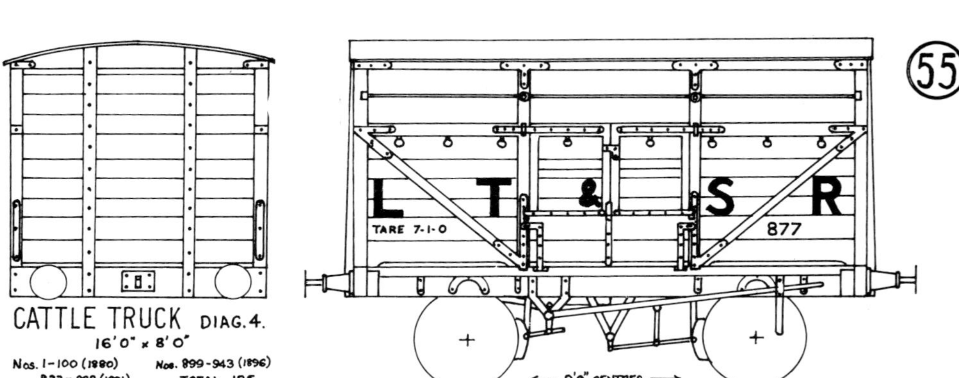

in the centre of the bottom plank on each side. For some reason, goods brake vans were classed as departmental stock.

Four of these vans were converted to ballast brakes at Plaistow in 1898 (Diagram 3). In their modified form they were numbered 1-4, but these need not necessarily have been their original numbers. At the same time a further 12 were rebuilt as open goods wagons, the sides being cut down to five planks, and the original ends retained. In this form they were renumbered 887-898 in the goods stock list. They could have been Nos. 5-16, but there is no confirmation of this, though it is logical to assume so, since they were the oldest vans in the fleet, dating from 1879/80.

Originally 46 brake vans were built between 1879 and 1885, all by the Metropolitan C&W Co.; probably because they all emanated from the same makers there is no mention of separate batches on the diagram. Since the total of goods brake vans was still 46 in 1912, the 16 conversions must have been replaced by new vans built at Plaistow from time to time, not necessarily built all at once.

Diagram 3 Ballast Brake Van (*figure 54*)

These four vans were altered from Diagram 2 at Plaistow in 1898. The modifications consisted of removing the bulkhead at one end, and boarding up the verandah at that same end. A small window was let into the top four planks at each side of the former verandah, and two larger windows were built into the end. Otherwise there was no difference between Diagrams 2 and 3. They were numbered 1-4, but there is no means of telling whether these were the original numbers, or a separate series.

Diagram 4 Cattle Truck (*figure 55*)

Another type of wagon of which a considerable number were built, though over a long period; they were the standard 16 ft in length and 8 feet in width, with 9 ft wheelbase. All the vans were of the same basic type, with some minor variations in the different batches, mostly in the arrangement of the brake gear. Most of the brakes were one side only, but a number were modified from time to time to have either side braking. The open sections at the top of each side had only one bar across them. Framing was outside, with an additional diagonal brace to the end sections. Four of them were converted at Plaistow to other uses, Nos. 874/5 to Meat Vans (Diagram 16) and 904/5 to Fish Vans (Diagram 13). Only one Midland number can be traced - 6434 - and this cannot be pinned down to any definite LT&SR number. There was some variation in lettering most had the standard large initials 'L.T.&S.R.', the '&' being on the left-hand door, not exactly in the centre, and the number on the next to bottom plank near the right-hand corner. A few vans, however, had small letters on the third plank from the top at the left-hand end, and the number in the corresponding place on the right. The vans were built as follows:-

Nos.	1 - 100	built	1880	by	Metropolitan C&W Co.
	833 - 882		1891		Midland C&W Co. *
	899 - 943		1896		Brown, Marshall & Co. +
			Total 195		

* Nos. 874/5 converted to Meat Vans, 1892 and 1899 respectively
+ Nos. 904/5 converted to Fish Vans, 1904

Diagram 6 Milk Van (*figure 56*)

These were six-wheeled, and of metal construction. There were only two, numbered 1 and 2, and built by Cravens Ltd in 1908. Though in the goods diagrams, they were rated for running in passenger trains, and as such were fitted with Westinghouse brakes; they were also painted in passenger livery. For some reason, possibly to bring the floor level down for loading purposes, they had only 2 ft 9 in. wheels, which ran in steel plate trunnions with long springs. The whole appearance of these vans was unusual, with their lowered solebars and heavy-duty buffers, sheet metal panels and wooden sliding doors with metal framing. There was a row of four louvres in the upper sides and ends, and 10 torpedo ventilators along the roof. Three doors were fitted in each side, sliding outside the main panelling, with a single short footstep beneath each. External stiffening of the metal panels was provided by vertical strips of 'T' iron. Lettering was in the small passenger style. Principal dimensions of the Milk Vans were: length 31 ft 6 in., width 8 feet, and wheelbase 22 ft 6in., equally divided.

Diagram 9 Covered Goods Vans (*figures 57 & 58*)

Next to the open wagons of Diagram 1, this was the most numerous type of wagon on the system, there being no less than 643 of them, built over a period of 17 years. The standard dimensions of 16 ft length 8 ft width, and 9 ft wheelbase were common to all. However, there were differences in the number of planks used in the sides and ends, the end planking often not corresponding with the sides. More wagons had 11 plank ends than any other number; the sides varied between 10 and 13. Unfortunately it is not possible to sort these out into batches, as too few photographs are available. Double doors were fitted, with an opening of six feet, and sliding inside the main panels. All the framing was outside, and the doors had slightly wider planks. There was some variation in the iron strapping; some vans of the

56

MILK VAN. DIAG. 6. 31' 6" × 8' 0". Nos. 1. 2. (1908). TOTAL 2.

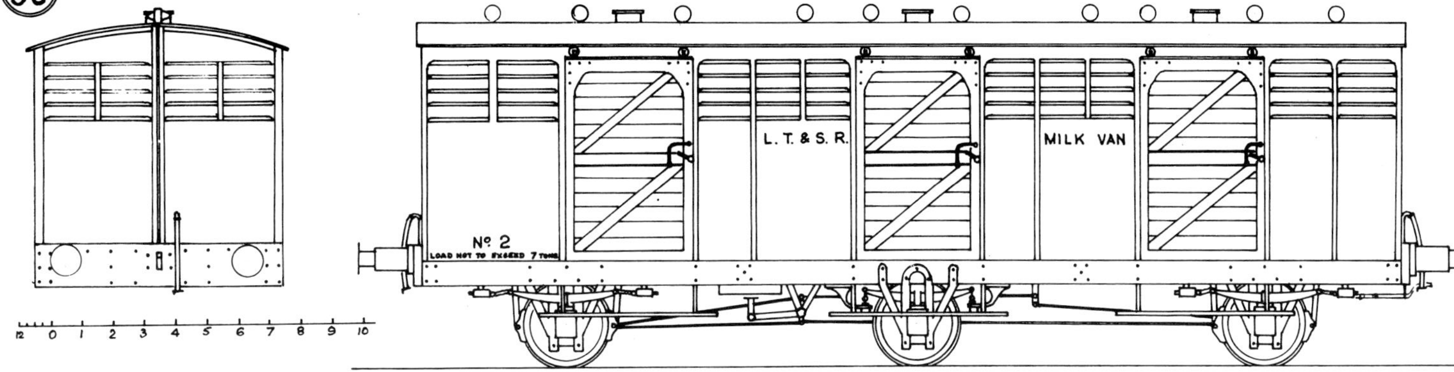

Plate 59: The very unusual six-wheeled vans built for milk traffic in 1908 to Diagram 6. There were only two of these, Nos. 1 and 2, mainly of metal construction, but with sliding doors of wood on metal frames. The most notable features are the low floor level and heavy-duty buffers.

Real Photographs

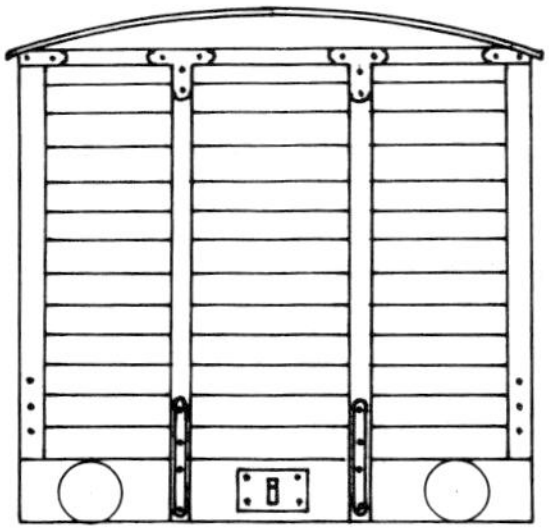

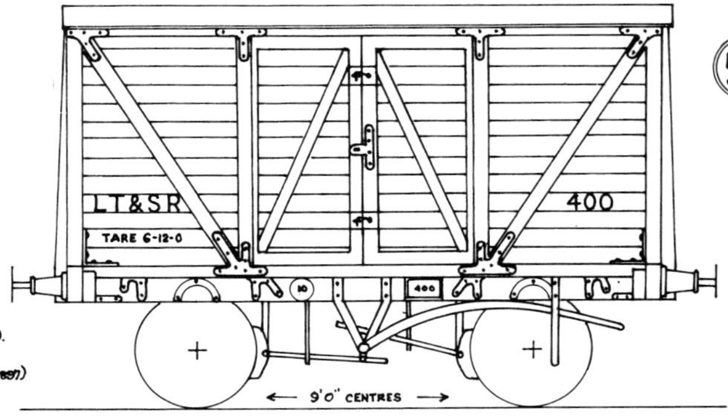

57

10 TON GOODS VAN DIAG 9.
16'0" × 8'0"
Nos. 323–572 (1885) 783–832 (1891) 944–1043 (1897)
TOTAL 400.

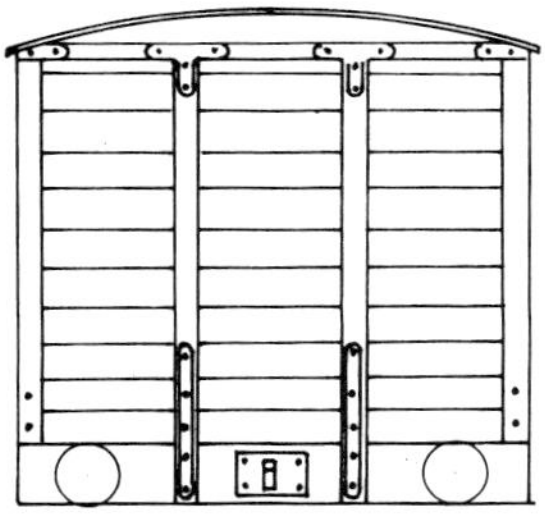

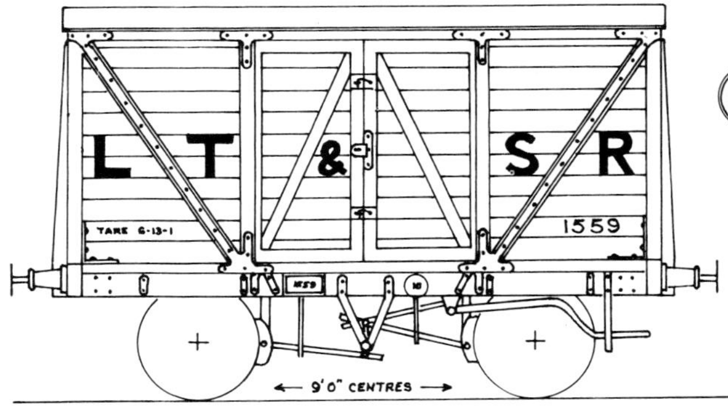

58

10 TON GOODS VAN DIAG. 9.
16'0" × 8'0"
Nos. 1200–1284 (1899) 1500–1649 (1902)
TOTAL 234

Plate 60: Single bolster wagon No. 1822 of Diagram 10, built in 1907; generally used in pairs. Note the peculiar shape of the hand brake lever. *Real Photographs*

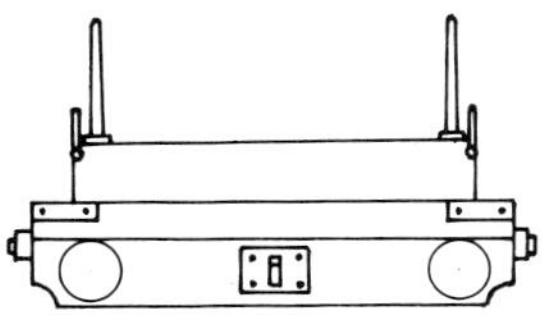

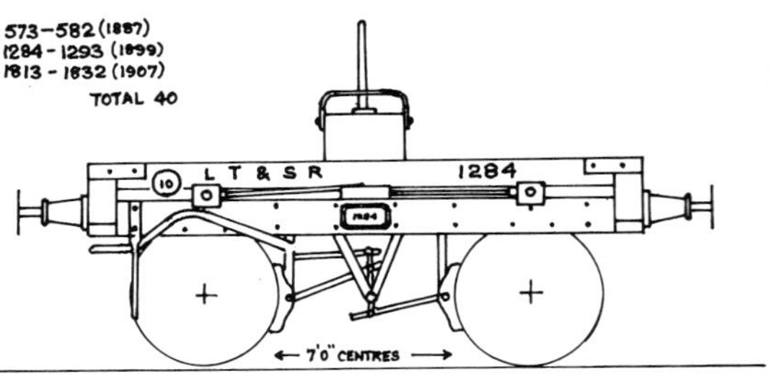

59

Nos. 573–582 (1887)
1284–1293 (1899)
1813–1832 (1907)
TOTAL 40

SINGLE BOLSTER WAGON. DIAG. 10
12'0" × 7'6"

1900-1902 batches having a continuous iron strap from top to bottom of the diagonal struts (not in the doors), while the earlier batches only had the peculiarly shaped corner plates. The brake gear also varied considerably, as can be seen from the two drawings. Earlier batches had brakes on one side only; some later ones had all four wheels braked, but hand levers on one side only. The covered vans (rated at 10 tons, as usual) were built as follows:-

Nos. 323 - 572	built	1885	by	Metropolitan C&W Co. *
783 - 832		1891		Oldbury C&W Co.
944 - 1043		1897		Metropolitan C&W Co.
1200 - 1283		1899		Metropolitan C&W Co.
1500 - 1649		1902		Metropolitan C&W Co. *

Total 643

* Nos. 323, 332, 1500-1513 were converted at Plaistow to meat vans (Diagram 16) in 1904

Diagram 10 Single Bolster Wagon (*figure 50*)

Also known as timber wagons, and often employed in pairs. There was not a great demand for timber wagons on the LT&SR, except for dealing with a certain amount of timber imported through Tilbury. These wagons, of which there were 40 in all, were of the standard pattern with short wheelbase (7 ft) and had a length of 12 ft. The single bolster in the centre could swivel, or could be locked rigid by means of a pin at each end. The load plate on these wagons said 10 tons, but on the Diagram they were rated at 8 tons only. Brakes were fitted to both sides, with the cranked hand levers at the same end. Some of these wagons were lettered in black.

Nos. 573 - 582	built	1887	by	Metropolitan C&W Co.
1284 - 1293		1899		Metropolitan C&W Co.
1813 - 1832		1907		Cravens Ltd

Total 40

Diagram 11 Double Bolster Wagon (*figure 60*)

This was a larger version of Diagram 10, 25 ft long and 8 ft wide, with a wheelbase of 15 ft. Only 10 of them were built, in 1904, by the Metropolitan C&W Co., and numbered 1775-1784. These too were rated at 10 tons. The brakes applied to one side only at first, but when revised brake regulations were imposed, they were fitted to both sides, with the levers at the same end. These wagons were mostly employed for carrying steel rails and girders; not

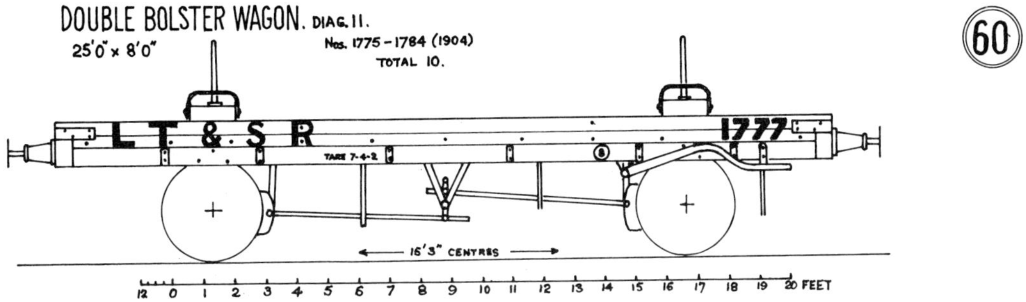

Plate 61: Double bolster wagon No. 1781, Diagram 11. This was the larger type of timber wagon, almost always used singly. The LT&SR had only 10 of these, Nos. 1775-1784, built in 1904. *Real Photographs*

61

10 TON BALLAST WAGON. DIAG.12
14'0" x 7'6"

Nos. 251–322 (1884)
1650–1674 (Secondhand, 1902)
TOTAL 97

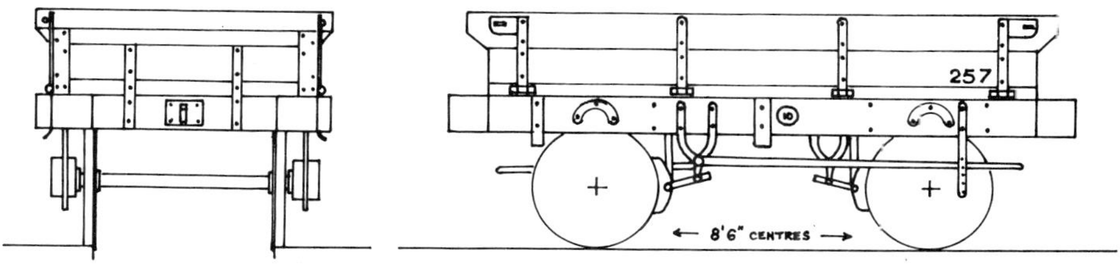

Plate 62: Diagram 12, ballast wagon. These had drop sides and dumb buffers. Forty (Nos. 251-290) were rebuilt as ordinary open goods wagons with spring buffers, in 1898. In 1902 twenty five ballast wagons (Nos. 1650-1674) were purchased secondhand in partial replacement. *Real Photographs*

10 TON OPEN WAGON DIAG 12A
Rebuilt from Ballast Wagon 1898.

62

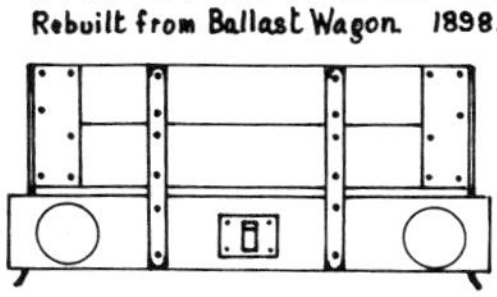

14'0" x 7'6"
Nos. 251–290 (Built 1884)
TOTAL 40

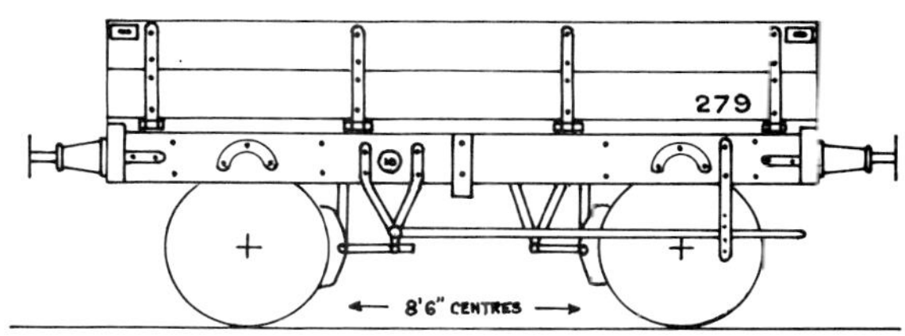

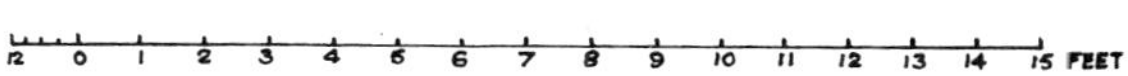

much in the timber trade required such length.

Diagram 12 Ballast Wagon (*figures 61 & 62*)

These were somewhat odd vehicles; first of all they were shorter than most other LT&SR open wagons, being 14 ft 6 in. long, 7 ft 6 in. wide, and with a wheelbase of 8 ft 6 in. they also had dumb buffers. The drop sides had two planks, and were 1 ft 10 in. high. There were two metal stops fixed on each side to prevent the sides dropping too far, but they were not equally spaced, one being in the centre and the other near the left hand end. After a few years an oblong metal plate was screwed to the side planks where they contacted the stops, as the latter were causing damage to the wood. There was only one brake block on each side of the wagon, applied by a very long hand lever fitted on one side only.

Seventy-two of these wagons, numbered 251-322, were supplied in 1884 by the Birmingham C&W Co. A further 25 were obtained second-hand from the same makers in 1902. These were numbered 1650-1674; there is no indication as to when they were actually built. Probably they were obtained to offset the loss of forty Ballast Wagons (Nos. 251-290) which were converted to normal goods wagons in 1898 at Plaistow. These 40 vehicles had spring buffers fitted and the ends modified; the second door stop at the left hand end was removed, leaving only the centre one, and the contact plates were also removed. No alteration was made to the brake gear (*see figure 39*).

In their original state the ballast wagons had 'Engineer's Department' in six inch letters on the top plank, and 'L.T.&S.R.' in standard 11 inch letters on the bottom plank. Those modified as goods wagons had the standard lettering moved upwards on to both planks, the joint going through the middle, and 'Engineer's Department' painted out.

Diagram 13 Fish Van (*figure 63*)

There were only two of these, nos. 904 and 905, modified in 1904 at Plaistow from standard Cattle Vans. They had the standard dimensions of 16 ft by 8 ft, with 9 ft wheelbase. Not a great deal of modification was necessary; the open sections at the top were filled in by louvre ventilators, and completely new doors were fitted. A lining and ice boxes were added inside. There is some doubt as to how they were painted; since they were not designed to run with passenger trains, the varnished teak livery was definitely not applied. It is alleged that they were painted in a light colour - possibly pale blue or pale green - with black lettering. However, as only one photograph exists of this type of van, it is possible that the livery shown was for photographic purposes only, and the vans may have gone into service in

Plate 63: Standard 10-ton open wagon, with five planks, goods Diagram No. 1. Over 800 of these were built between 1876 and 1904. Minor variations were apparent in different batches. No. 1748 was one of the last batch of 100 built in 1904. *Real Photographs*

63

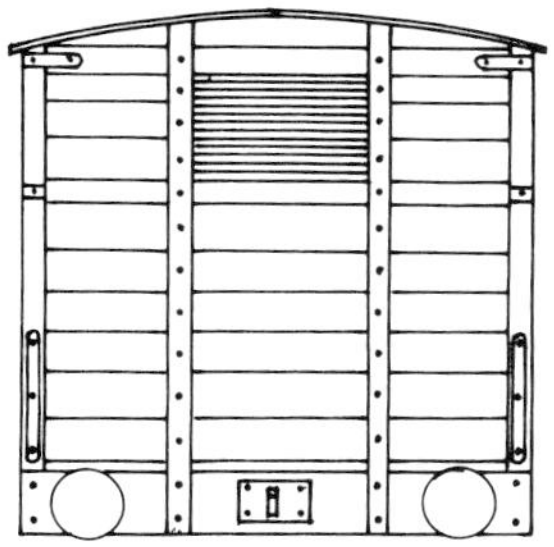

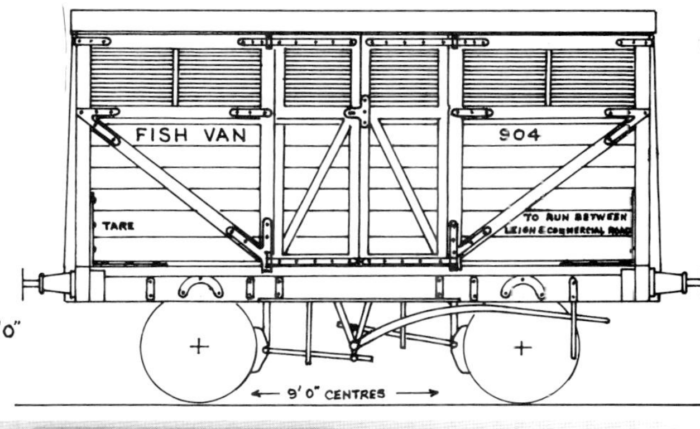

FISH VAN DIAG. 13. 16'0"x8'0"
Rebuilt from Cattle Truck 1904.
2 ONLY. Nos. 904, 905.

Plate 64: There were only two of these fish vans (Diagram 13), both rebuilt from cattle vans of Diagram 4 in 1904. Unlike the practice on most other railways, they were not fitted with automatic brakes, and were not run attached to passenger trains.
Real Photographs

64

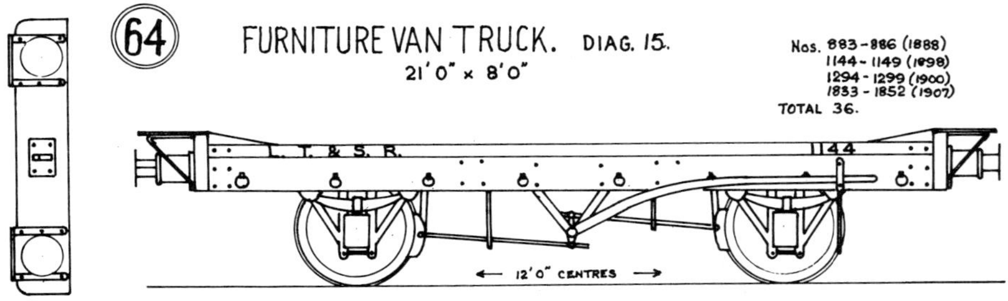

Plate 65: Diagram 15, furniture van truck No. 1841, built in 1907. These were a cross between a well wagon and an open carriage truck, having a shallow well in the floor to bring the overall height down to gauge limits when loaded with a standard pantechnicon. *Real Photographs*

65

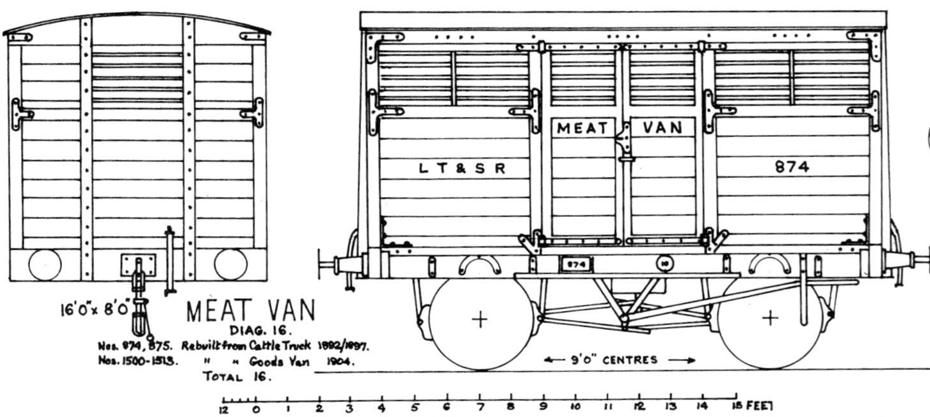

66

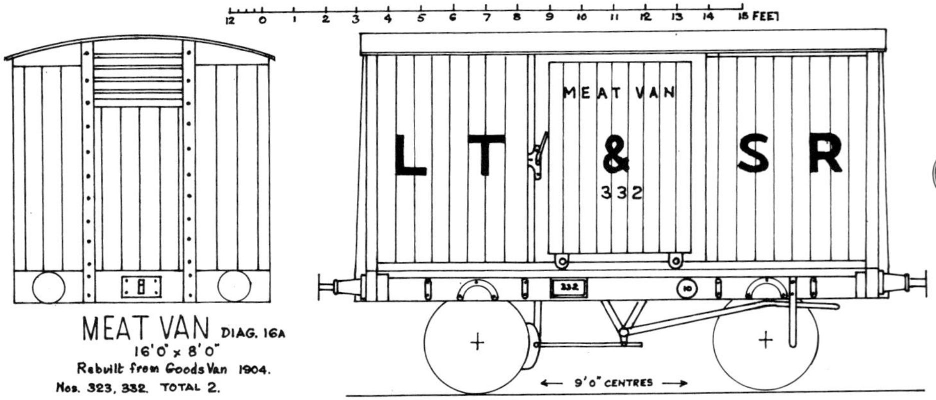

the standard grey paint. This is a point which is incapable of a satisfactory conclusion. They were never Westinghouse-fitted, which is unusual, for on most other railways fish vans were almost always attached to passenger trains.

Diagram 15 Furniture Van Truck (*figure 64*)

These were a kind of cross between an open carriage truck and a well wagon, since they had some features of both in their design. The unusually deep solebars - 15 inches - will be noted. This was on account of a nine inch well in the floor, designed to bring down the overall height when loaded with a standard pantechnicon of the period. It was no easy matter unloading these wagons, since the vehicle being carried had to be hauled up a ramp at the end of the wagon to allow it to pass over the buffer beam. Usually a capstan had to be employed. Steel running plates were fitted over each buffer. There were eight rings fitted in the lower part of the solebars on each side through which securing ropes for the load could be passed. The last batch, built by Cravens, had only seven rings. Load rating was the usual 10 tons. These wagons were 21 ft long and 8 ft wide, with a wheelbase of 12 ft. Braking as usual was by one block on each side, with hand lever on one side only. Apparently there was not a great demand for these wagons at first, for they had a very slow start, in batches of four or six, but suddenly in 1907 twenty of them were acquired.

Nos. 833 - 886	built	1888	by	Metropolitan C&W Co.
1144 - 1149		1899		Metropolitan C&W Co.
1833 - 1852		1907		Cravens Ltd

Total 36

Diagram 16 Meat Van (*figures 65 & 66*)

None of these were built new for the purpose of carrying meat, they were all rebuilt from other types of vans. Two were originally Cattle Vans, and the rest were standard Goods Vans. Though the LT&SR Diagram gives no dimensions, it can be inferred from the original types that they were all of the standard measurements, 16 ft long, 8 ft wide, and with a wheelbase of 9 ft. The first two converted, Nos. 323 and 332, were totally different from the rest, having vertically boarded sides without louvres, inside framing, and a single sliding door on each side, outside the panelling. There is some argument as to whether both these vans were converted, or only one of them; the Diagram gives only one number, 323, but on the other hand the only available photograph clearly shows the clearly shows the number as 332. This argument is incapable of solution, since no more information is available.

Plate 66: Standard gunpowder van no. 1800, Diagram 17. A standard RCH all-metal van, one of 25 built in 1904, chiefly for supplies of ammunition etc. to the military establishment at Shoeburyness. They were painted bright red, with white roofs and lettering.

Real Photographs

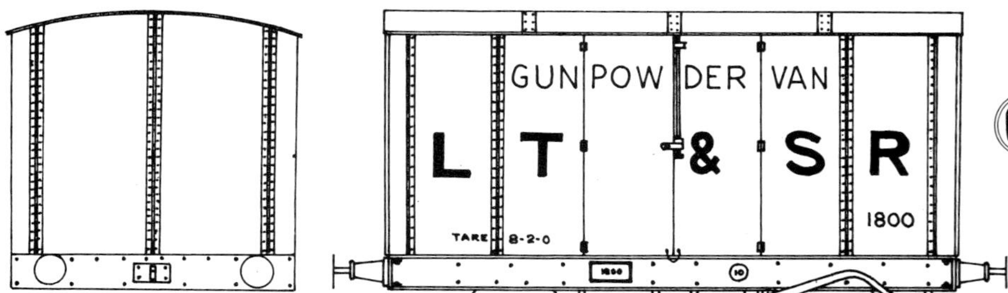

EXPLOSIVES VAN DIAG. 17
16'0" x 8'0"
Nos. 1787 - 1811 (1904) TOTAL 25

Plate 67: The standard meat van, Diagram 16. None of these were built new as meat vans, Nos. 874/5 were originally cattle vans of Diagram 4, while the other 16 Nos. 323, 332 and 1500-1513 were all rebuilt from standard covered goods vans of Diagram 9.

Real Photographs

The author inclines to the theory that both of them were converted, if only on the assumption that a total of 18 vehicles is much more likely than 17; the LT&SR did not seem to favour odd numbers of vehicles, except in small quantities - one, three, or five. Nos. 323 and 332 were apparently not Westinghouse-fitted, unlike the rest, though they carried passenger livery. They also had hand brakes on only one wheel.

Nos. 874 and 875 were converted from Cattle Wagons in 1892 and 1899 respectively. The resulting vehicles resembled the Fish Vans very closely, the open sections at the top being filled in by louvres, and double hinged doors on each side; the diagonal struts in the outside framing were omitted. The planking was horizontal. In 1904 fourteen more were converted, Nos. 1500-1513, from standard goods vans of Diagram 9, to the same design as Nos. 874/5. All these had Westinghouse brakes, and were painted in the passenger livery. Internally they had a separate lining all round with a cavity between it and the outer panels (as in the Fish Vans) into which ice could be packed. Drainage holes in the bottom allowed water to run away. Large meat hooks were suspended from the inside of the roof on steel rails. They were used for the meat traffic between Thames Haven or Tilbury and Commercial Road warehouse in Whitechapel, sometimes run as separate trains, or in twos or threes attached to passenger trains.

Diagram 17 Gunpowder Van (*figure 67*)

It seems rather curious that a small railway like the LT&SR should require as many as 25 Gunpowder Vans, far more than some of the larger companies, but the only logical conclusion is that they were necessary to cope with the requirements of the Shoeburyness garrison. The batch were all built at the same time in 1904 by the Metropolitan C&W Co., and carried the numbers 1787-1811. The dimensions were the usual standard. Bodies were constructed to RCH specifications, all steel, with steel roof, double hinged doors, and 'T' strapping over the joints in the plating. These vans were painted red, with white lettering. Either-side hand brakes were fitted. For some reason, the complete Midland renumbering of the gunpowder vans has been recorded:- Nos. 1787-1811 became Midland 109985-110000, 115053, 117298-117300, 117501-117505 respectively.

Diagram 18 Bogie Well Wagon (*figure 68*)

This was a special wagon designed for exceptional loads; only two were built, Nos. 1785 and 1786, by the Metropolitan C&W, Co. in 1904. They were 8 ft wide, and 51 ft 4 in. long over headstocks; they ran on a pair of four-wheeled bogies of 5 ft 6 in. wheelbase, fixed at 40 ft 4 in. centres. The floor-

68

20 TON
BOGIE WELL WAGON.
DIAG. 18.
51' 6" x 8' 0"

Nos. 1785, 1786. (1904)
TOTAL 2.

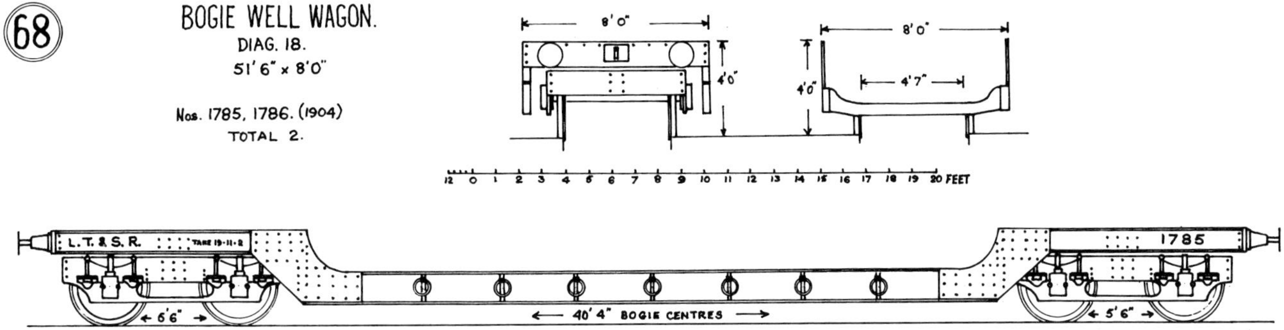

Plate 68: Heavy duty bogie well wagons, Nos. 1785/6 of Diagram 18. Rated at 20 tons, they were somewhat restricted as to use if booked to go off LT&SR metals. Used for transporting heavy castings, occasionally also large gun barrels. *Real Photographs*

ing was sheet metal at the ends, over the bogies, and timber baulks in the central well. Construction was chiefly of channel steel, the 30 ft-long girders forming the well were fastened to the raised girders at each end by heavy shaped steel gusset plates, one inch thick. Clearance above rails in the well was 10½ inches. The bogie frames were slotted out from sheet steel, and had heavy duty springs and axle boxes. The wagons were rated at 20 tons, and the minimum curve they would negotiate was 1¾ chains. All wheels were braked, the brakes being applied by a handwheel at the extreme right-hand of the frames on each side. Restrictions on use were given as 'Not to be sent to Poplar (Mid.) or Millwall Docks without consulting Chief Goods Manager' also 'Assent of Great Eastern Railway required before loading to or over GER.'

Diagram 19 Fire Engine Wagon (*figure 69*)

This is something of an enigma. By its appearance, it was built at the same time as the fire engine, namely 1883, but its number, 1853, dates from 1907. There is some evidence for it not having carried a number at all for most of its life, which would validate 1907 as the date when it was first given a number. On the other hand, it may have had a departmental number until 1907; there is no means of ascertaining which of these two statements is correct - or at least, nearest to the truth. In 1883 the LT&SR purchased a small steam-operated fire engine from Shand, Mason & Co. of London, which was suitable for conveyance by a railway wagon to any location where it might be required. The necessary wagon was built, but where, is not recorded. It was a two-plank open wagon with drop sides and open ends, and was only 12 ft long and 7 ft 11 in. wide, with a 7 ft wheelbase. It was fitted with run-off plates over the buffers for loading and unloading the fire engine. Braking was the usual one block on each side, applied by a curved hand lever on one side only. Lettering was in the standard small type in white on the bottom plank. Screw couplings were fitted, but not the Westinghouse brake.

This completes the working goods stock, but in addition there were eight vehicles which did not appear in the Diagram Book, and were attached to the Locomotive Department to make up the breakdown train, kept at Plaistow. First there was the original breakdown crane, a 10-ton handworked one, built in 1881 by H.J. Ellis & Co., of Manchester. It was constructed on a six-wheeled base, 18 ft long and 7 ft 6 in. wide, having an equally-divided wheelbase of 12 ft this being for all practical purposes a normal wagon underframe. Two footsteps of locomotive type were provided, at the left-hand corner of each side, and at the same corners a hand wheel for applying the brake was fitted. The brakes were fitted to four wheels only, including the centre pair.

Plate 69: A special truck (Diagram 19) was provided for transporting the small Shand, Mason fire engine, originally purchased in 1883. Though not specifically stated, it is thought that the truck was of the same vintage, although the goods stock number allocated (1853) dates only from 1907. However the truck is obviously of much earlier date. *British Rail*

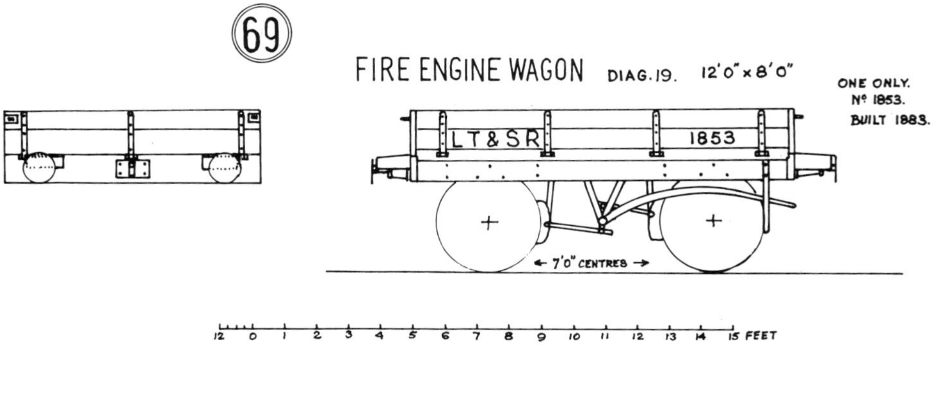

The superstructure of the crane was mounted on a raised circular base 4 ft in diameter, the centre of which was mid way between the centre and end axles. On this was mounted a pair of cast iron girders of special shape, 14 ft 6 in. long, which formed the crane base. At the forward end the jib was mounted, this being a hexagonal wooden pole 18 ft 6 in. long and one foot in diameter. At the other end of the base was an oblong moveable counterweight mounted on four small wheels, which could be clamped in any desired position on a pair of rails. The operating handle of the crane was on the right hand side (facing towards the jib) working through a small pinion onto a large gear wheel attached to the lifting barrel, lifting being done by a chain. A second pinion on the other end of the handle shaft, and controlled by a dog clutch, meshed with another large gear wheel operating a smaller barrel which served to raise and lower the jib. This was fitted with a band brake and a ratchet and pawl to maintain the height of the jib in any desired position. Swivelling of the crane on its base truck was done by hand, not mechanically.

The crane was accompanied by a runner wagon (or Match Truck) apparently built by the crane makers at the same time. This was 14 ft long and 7 ft 6 in. wide, running on four wheels spaced at 9 ft centres. This was a normal flat wagon, with hand brake on one wheel only, having a support for the crane jib built up in the centre. Around this was a large wooden box structure with sloping lids on either side to carry tools, spare chains, and other necessary gear. This box was 12 ft long, 6 ft wide, and 1 ft 3 in. high at the sides, and was rather plentifully supplied with iron strapping. The hand crane and runner coped with all the incidents on the system until 1905, when a 15 ton steam operated crane was obtained. The hand crane was then used for minor incidents and various engineering purposes which did not warrant the calling out of the larger crane. Neither the hand crane nor the runner appear to have been allotted a stock number (*figure 75*).

15-Ton Breakdown Crane (*figure 70*)

Cowans, Sheldon & Co. of Carlisle provided a 15-ton breakdown crane (Works No. 2933) in 1905. It was the last example of this particular design to be built for a British railway, though a number were made subsequently for railways in Ireland and overseas. It was constructed on an eight-wheeled base, the two axles under the machinery end being rigid with the main frames, while the jib end was fitted with an Adams type bogie. The side frames, of steel, edged with angle, were two feet deep, and over buffer beams were 23 ft 6 in. long, with an overall width of 7 ft. 11½ in. The wheelbase was 6 ft 3 in. (bogie) + 5 ft 4 in. + 6 ft 6 in., total 18 ft 1 in. The rigid pair of wheels had outside bearings, but the bogie bearings and frames were inside. Brakes

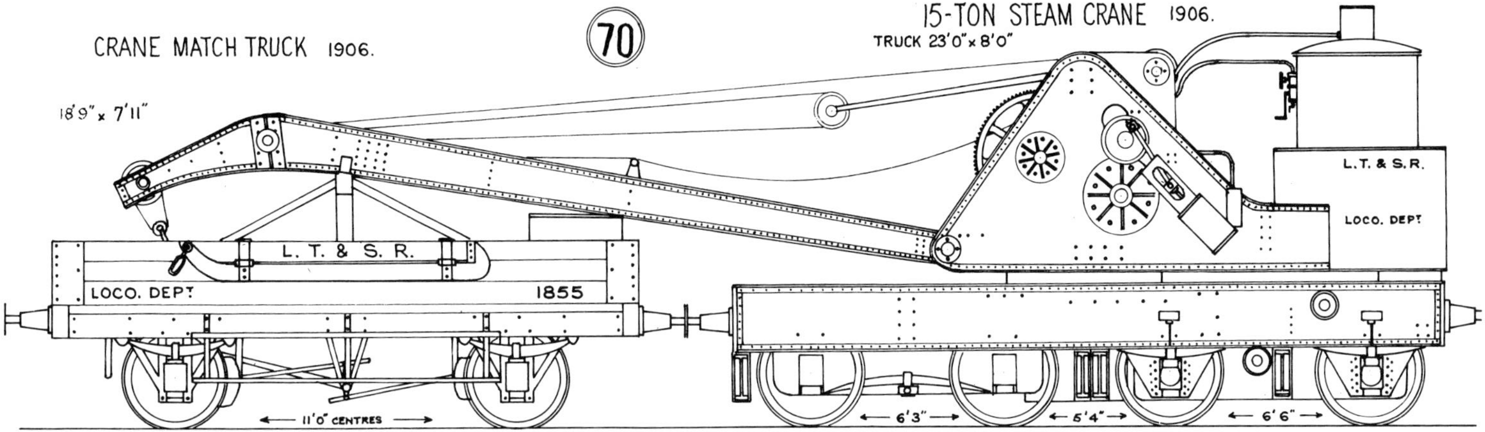

Plate 70: The 15-ton steam breakdown crane built by Cowans, Sheldon & Co. of Carlisle in 1905. It remained at Plaistow until 1933, when it was transferred to the West Midlands, and was broken up in 1963. *British Rail*

on the rigid wheels were applied by a hand wheel on either side placed between the axles. There were three boxes for extendable outriggers, one at the extreme jib end of the carriage, one between bogie and rigid wheels, and the other close to the end rigid axle.

The side frames of the crane itself were roughly triangular plate, with angle edging, fixed to a circular roller race five feet in diameter in the centre of the carriage. Attached to an extension of the side plates at the rear was a vertical boiler with coal bunkers on each side. The top of the boiler was a dished plate which carried in its centre the chimney. This was short to clear the loading gauge when in transit, but for working purposes a two feet extension was provided. On either side of the main plates was fixed a cylinder, 7 in. x 12 in. at an angle of 45°, which drove a main shaft from which, by means of clutches and gears, all the movements of the crane were achieved. The jib was of the so-called 'swan neck' pattern, built up from steel plate and angle. All hoisting movements were done by chains.

The runner, which was built at Plaistow, was of a different type to that provided for the hand crane. It was 18 ft 9 in. long and 7 ft 11 in. wide overall, with four wheels at 11 ft centres, and mainly of metal construction. The sides were two feet high, and in the middle was a strong triangular wooden structure four feet high to support the crane jib when travelling. Two pieces of special lifting gear were slung outside from the top edges of the truck, and other equipment was carried inside. Two long footboards were fitted, one at axle level and the other at the bottom of the solebars. Both the runner and the crane were fitted with screw coupling and Westinghouse brake gear, for use in transit only, for under working conditions without an engine attached, the air brake was inoperative. The crane had no number, only 'L.T.&S.R.' and 'Loco. Dept.' on the bunker sides. However, the runner carried the same lettering in larger characters, but with the addition of the number 1855, which was in the normal goods series.

The crane remained at Plaistow all through the Midland regime, and into the LMS period, being then numbered MP 37. In 1933 it was transferred to Bescot, in the Midlands, and renumbered RS 1024. This was its last posting, and it was broken up in 1963.

Stores Wagon (*figure 74*)

This was another open wagon of rather unusual design, built at Plaistow. For most of its length it had two planks, but for a short length at each end it had a third plank, tapered down to the main height, which was 1 ft 4 in. the end height being 2 ft. It was 25 ft 6 in. long and 8 ft wide, running on four wheels at 17 ft centres. In the middle, between the axles, was a low-slung storage box, 13 ft long and 2 ft 3 in. deep, to which access was by two lift-up doors on each side, hinged to the bottom of the solebars. Screw couplings were fitted, but not Westinghouse brakes. The hand brake, operating on one wheel on each side, was applied by a small wheel near the right-hand end. This wagon was used to carry a small emergency bogie and various other spare parts, wood blocks, jacks, and so forth. It was usually the fourth vehicle in the train, being marshalled immediately in front of the two vans, which were always in the rear. This wagon was numbered 1854 in the goods stock. As shown in the drawing it was lettered 'Break Down Train' (in three words) on the upper plank, and 'L.T.&S.R.' on the lower plank at the left-hand end, with the number at the opposite end. In the centre of the bottom plank it was lettered 'Loco. Dept.' in small characters.

Packing Wagon (*figure 74*)

This was the third wagon in the train, marshalled immediately behind the crane and runner. A smaller open wagon than No. 1854, it was 16 ft long and 8 ft wide, with a 10 ft wheelbase. It was also a two-plank wagon, with sides 1 ft 4 in. high. There was nothing unusual about this vehicle, it was similar to the ordinary open goods stock. It was lettered similarly to no. 1854, with 'Break Down Train' in three words on the top plank, 'L.T.&S.R.' in the left-hand corner of the lower plank, 'Loco. Dept.' in the centre, and 'No. 3' in the right-hand corner. Why this particular wagon carried a Loco. Dept. number, and not a goods number like the other two, is not explained.

Tool Van No. 1 (*figures 71 & 72*)

There were two Tool Vans, not alike, though both were rebuilt at Plaistow at the same time from standard Goods Brake vans of Diagram 2. Loco. Dept. No. 1 was very similar to the ballast brakes (Diagram 3) but had the verandah at one end shortened by extending the side planks by six inches. Two glass ventilator lights were let into each side just under the roof. The end of the open verandah had a gangway cut out of the middle, four feet wide, with a fall plate by which access could be had to the other van while the train was in motion. The hand brake column was moved from inside the van onto the verandah, and the vehicle retained its full length footboard and side handrail.

Tool Van No. 2 (*figures 71 & 72*)

Also rebuilt from a goods brake van, but in this case the verandah was lengthened and the lower half boarded in, so that the only means of access was through the end, from Van No. 1. This van also retained the footboard,

Plate 71: The two specially modified goods brake vans for use in the Plaistow break-down train. Both were rebuilds of standard brake vans (Diagram 2) and were kept close coupled. The adjoining ends were modified with an opening and fall plates to facilitate passing from one to the other while in motion. No Diagram was issued for them. *Real Photographs*

TOOL VANS. (REBUILT FROM OLD GOODS BRAKES 1906.)

71

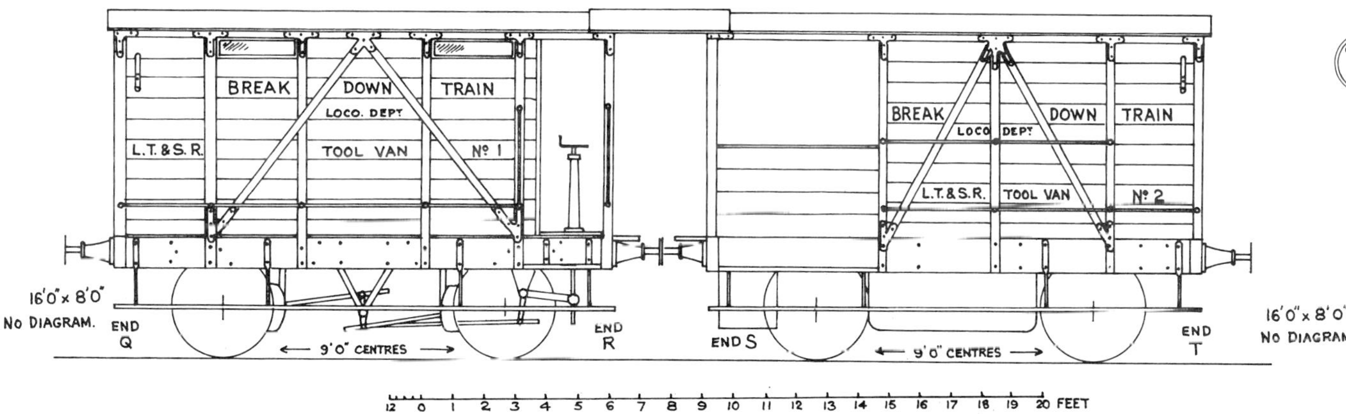

72

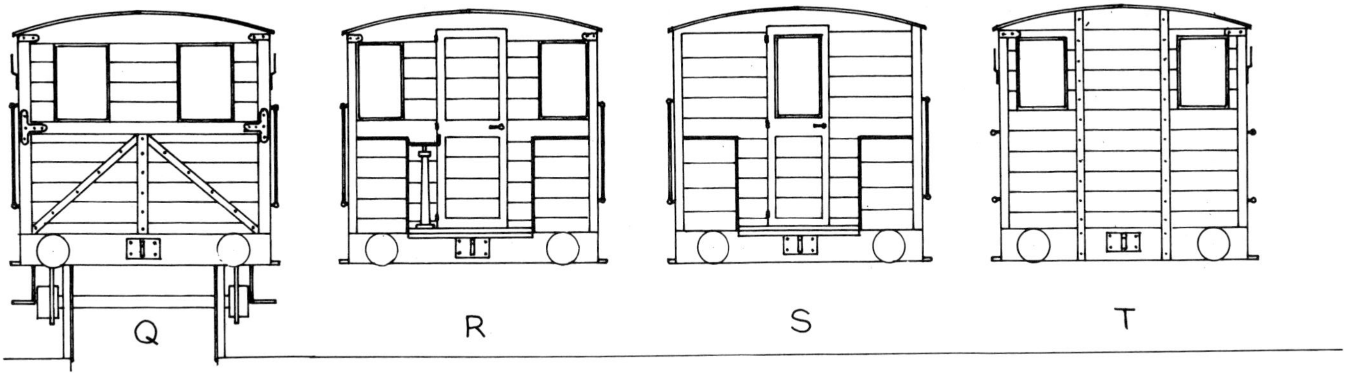

73

STORES TRUCK.

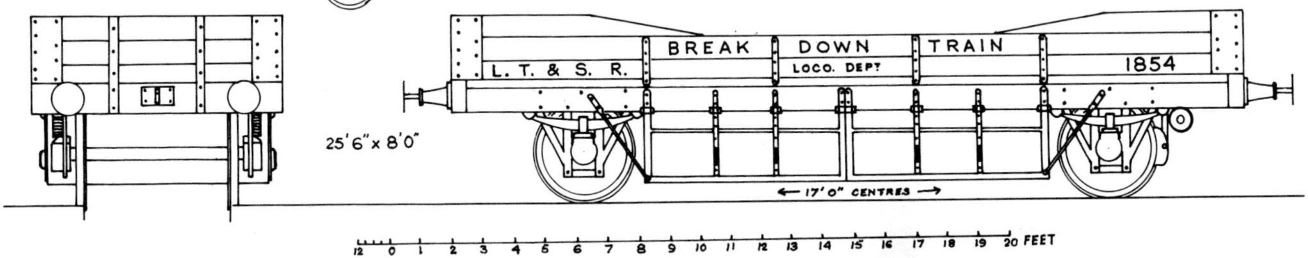

74

PACKING WAGON.

Plate 72: Included in the breakdown train was this packing wagon, for carrying wooden blocks, old sleepers, etc. for use when jacking up derailed vehicles, and supporting the outriggers of the steam crane. *British Rail*

Plate 73: The breakdown train stores wagon, which carried an emergency bogie, tools, chains, ropes, and other odds and ends of equipment which might be required. A large storage box was slung underneath the wagon, between the axles. For some reason, it was numbered in the goods stock list. *British Rail*

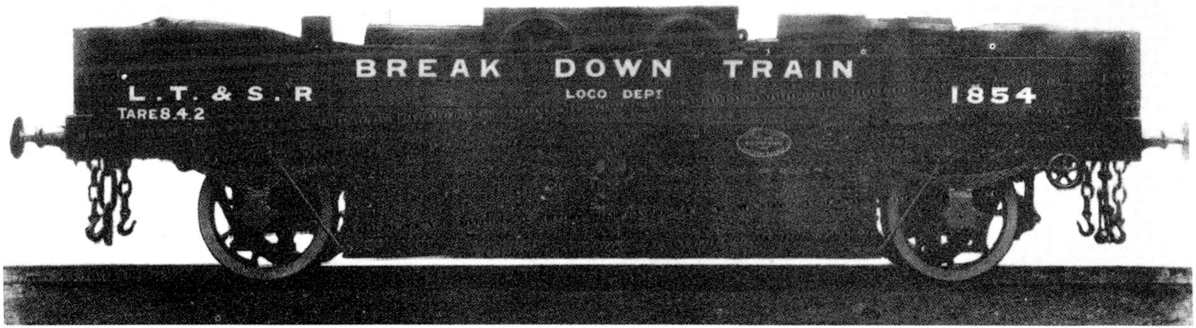

but had an extra lengthwise handrail fitted two feet above the existing one. The space between the roofs of the two vans was covered by an extra piece of roof fixed to van No. 1, but free to slide over the roof of No. 2. Both vans were lettered in the same way, though the positioning of the lower part was slightly different.

It has been alleged that the Breakdown Train, including crane, was painted in locomotive green, but this cannot be confirmed; though it is a feasible assumption if taken from photographs, for they certainly indicate a darker colour than the standard goods grey. However, the author fears that the reader will have to make up his own mind on this subject, for no amount of digging into documents has produced any comment at all on the livery of the Breakdown Train. It can be taken for certain that the lettering was the standard off white.

The line was singularly free from serious accidents. Only one fatality was said to have been caused, as a result of a collision at Barking on 20th September, 1875. On this occasion one of a series of five excursion trains returning from Southend ran into the back of the previous train, killing one passenger. The BOT inspector gave as his opinion - the cause of the accident was lack of care on the part of the driver, but at the same time pointed out that to send out an 18 coach train with only one guard to attend to the brakes was the height of folly. (At that time the trains were fitted with Clarke & Webb chain brakes.)

A less notable accident took place on 17th August, 1859, when an express from Thames Haven took the curves at Tilbury at too high a speed, and was partially derailed, with no serious injury.

Minor collisions took place at Fenchurch Street on 1st August, 1859, and at Abbey Mills Junction in February 1879, but with very little damage and no casualties. It will be noted that all these took place before the LT&SR had taken over the working of the line. The only accident accountable to the LT&SR in its own right was the derailment at Pitsea in December 1897, when a train took the curve from the Tilbury line too fast. Though three coaches were damaged badly enough to warrant their being broken up, there were no casualties and only minor injuries.

A much more serious event occurred in July 1870 when a carriage shed at Tilbury was destroyed by fire, including, twelve brand new coaches which had just been delivered. This may possibly have been the cause of the aberrations in the stock list, and some renumbering, in the early years.

What used to be the LT&SR has suffered much less in proportion to its size than most other railways in the way of closures during the past thirty years. The entire main line from Fenchurch Street and St Pancras to Shoeburyness is still in full use; the Romford-Ockendon-Grays section, the Tilbury-Pitsea line and the Thames Haven branch have gone but these were of minor importance, and were only a small proportion of the system. The intensive commuter traffic to and from Southend ensures its survival for many years to come. It even had a special class of locomotives built for it - the Stanier three-cylinder 2-6-4 tanks, and it was the guinea pig for the successful trials of the Strowger-Hudd system of automatic train control. The line is in no danger yet, but it is a pity that the interesting and picturesque locomotive names have been lost. What could replace being hauled from Fenchurch Street to Southend by *Commercial Road*.

Plate 74: An ex-LT&SR Vacuum Cleaning Van built in 1912 seen here in the British Railways era at Windermere on 22nd July, 1966 as Mobile Charging Plant No. 31.

H.C. Casserley

10 TON HAND CRANE & RUNNER. NO DIAGRAM.

NOT NUMBERED. ONE ONLY. BUILT 1881.

75

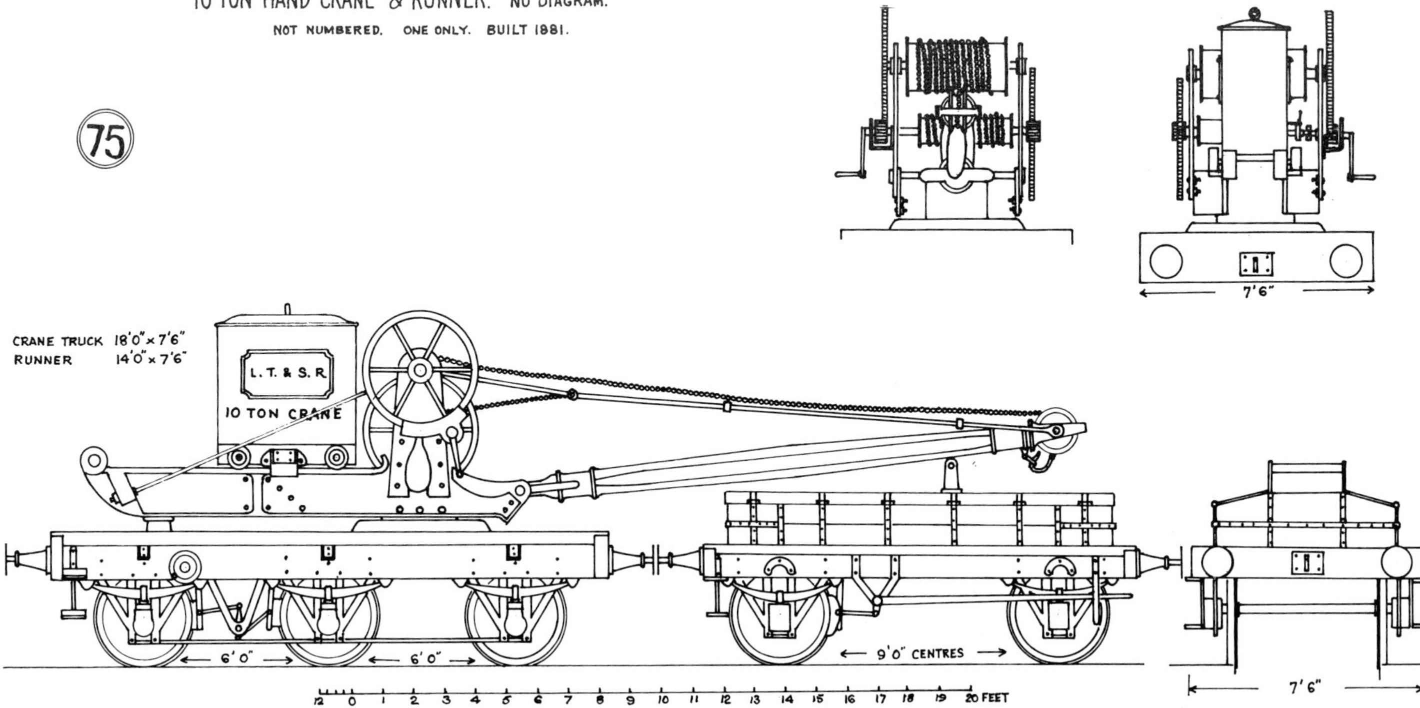

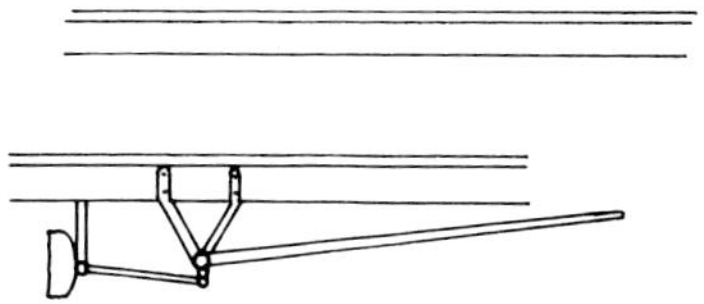

ONE WHEEL ONLY.

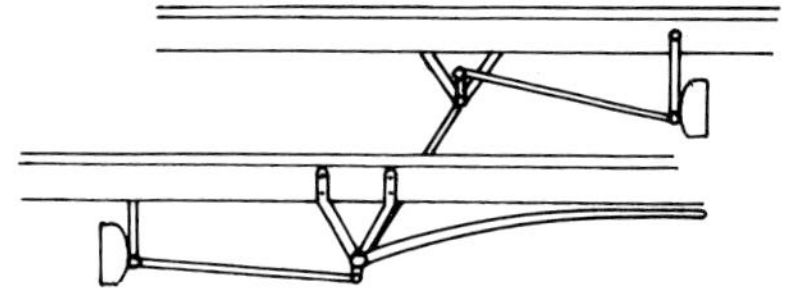

ONE WHEEL EACH SIDE. ("CROSS CORNER" BRAKES.)
LEVER ONE SIDE ONLY.

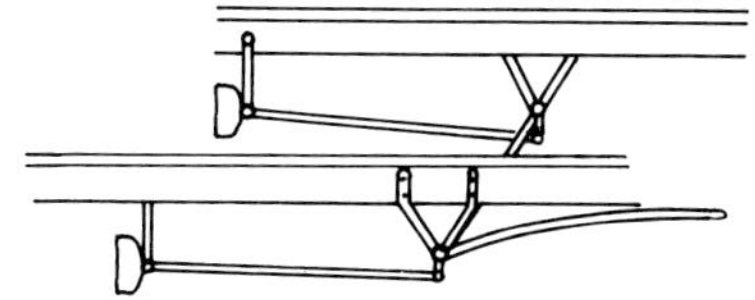

ONE SIDE ONLY.
BOTH BRAKE BLOCKS ON SAME AXLE.

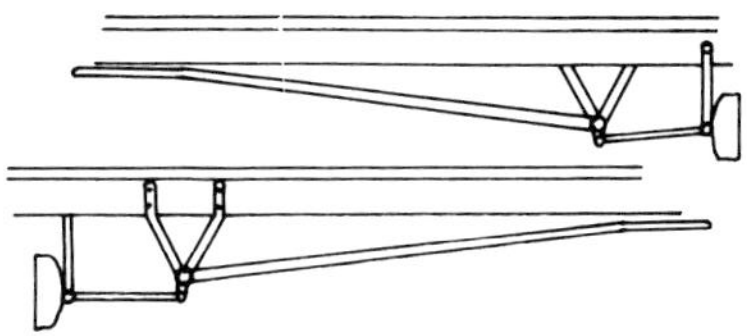

EITHER SIDE BRAKES.
(INDEPENDENT)

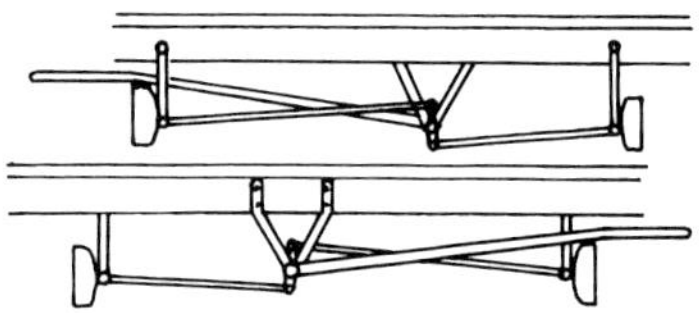

EITHER SIDE BRAKES.
(INDEPENDENT – BLOCKS ON ALL WHEELS.)

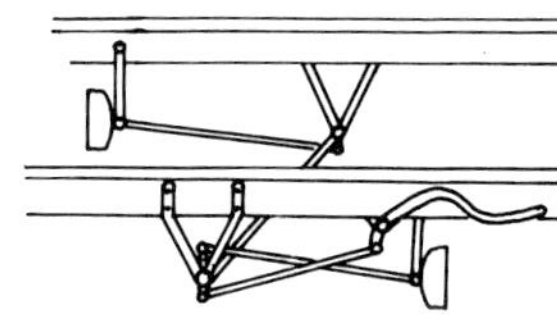

"CROSS CORNER" BRAKES.
SHORT LEVER TYPE – ONE SIDE ONLY.

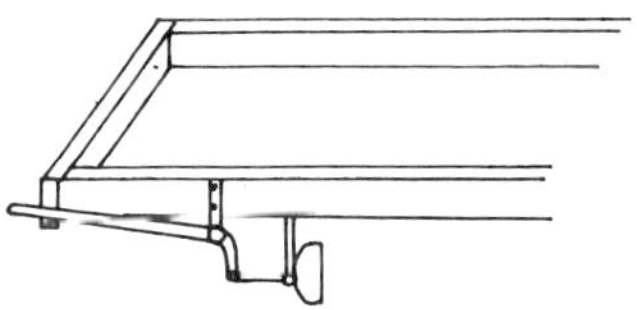

"CROSS CORNER" BRAKES
SPECIAL SHORT LEVER TYPE.

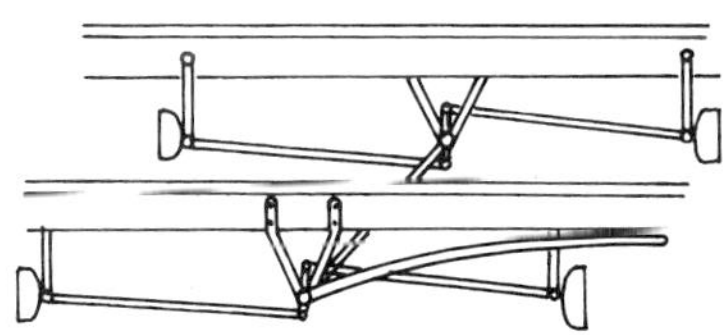

ONE SIDE ONLY.
BRAKE BLOCKS ON ALL FOUR WHEELS.

GOODS WAGON
BRAKE GEAR ARRANGEMENTS.

DIAGRAMATIC ONLY.
NOT TO SCALE.

Appendix One

Passenger Diagram List, 1912

Dia.	*Wheels*	*Type*	*Length*	*W'base*	*Dates*	*Compts*	
1	4	First	26′ 0″	15′ 3″	1876-1878	4	
2	4	First	26′ 0″	15′ 0″	1876-1886	4	
3	4	First	24′ 8″	14′ 0″	1883?	4	(a)
4	6	First	29′ 9½″				
5							
6	8	First	48′ 0″		1901-1911	7	(c)
7	4	Composite	24′ 9″	14′ 0″	1876	2x1, 2x3	
8	6	Composite	29′ 9½″			3x2, 2x3	
9	6	Composite	29′ 9½″			3x1, 2x3	
10	8	Composite	48′ 0″		1901-1911	2x1, 6x3	
11	4	Third	26′ 0″	15′ 3″	1876-1877	5	
12	4	Third	25′ 9″		1877-1886	5	
13	4	Third	28′ 3″	16′ 3″	1878-1883	5	
14	4	Third	28′ 4½″	16′ 0″	1886	5	
15	4	Third	24′ 8″	14′ 0″	1877	5	(b)
16							
17	6	Third	33′ 11″			6	
18	8	Third	46′ 0″		1901-1911	8	(c)
19	4	Bk. Third	24′ 9″	14′ 0″	1877		(d)
20	4	Bk. Third	25′ 9″	15′ 0″	1878	2	(d)
21	4	Bk. Third	26′ 0″		1877-1883	2	(e)
22	6	Bk. Third	33′ 11″			3	(d)
23	6	Bk. Third	31′ 6″				
24							
25	8	Bk. Third	46′ 0″		1901-1911	4	(d)
26	8	Bk. Third	46′ 0″		1906-1911	4	(e)
27	4	Saloon	24′ 9″	14′ 0″	1877		
28	6	Saloon	34′ 0″				
29	4	Brake Van	24′ 9″	14′ 0″	1877-1886	-	
30	4	Match Wagon	16′ 0″	10′ 0″	1907	-	
31	4	Horse Box	16′ 0″	10′ 0″	1878	-	
32	4	Horse Box	17′ 6″	10′ 0″	1902	-	
33	4	Carr. Truck	15′ 6″	10′ 0″	1878	-	
34	4	Luggage Van	16′ 0″	9′ 0″	1896	-	
35	4	Bullion Van	15′ 7″	9′ 6″	1911	-	
36	6	Milk Van	31′ 6″		1908	-	(f)
37	8	Corr. Compo.	47′ 6″		1912	2	(g)
		Corr. Third	47′ 6″		1912	2	(g)
		Corr. Bk. Third	47′ 6″		1912	3	(g)
?37	4	Vacuum Cleaner		25′ 0″	1912	-	
37A	4	Loco. Stores	25′ 0″		1916	-	(h)

38	6	Scenery Van	34′ 0″	21′ 0″		-	
39	8	First	52′ 9½″		1912	8	(k)
40	8	Third	52′ 9½″		1912	10	(k)
41	8	Third	52′ 9½″		1912	9	(k)
42	8	Bk. Third	52′ 9½″		1912	7	(k)
43	8	Saloon	46′ 9″		1912		
44	8	Electric Coaches					(l)

(a) One only (no. 17).
(b) One only (No. 236). This and (a) above must have been renumbered, but no other trace can be found.
(c) Lavatory-fitted coaches included on same diagram.
(d) Guard's van in centre.
(e) Guard's van at one end.
(f) Was also included in the goods diagram list. (Diagram 6)
(g) The Ealing corridor stock. All three varieties of coaches were apparently on the same diagram.
(h) May have been rebuilt from a Vacuum Cleaner Van, or may have been built new in this form. The available information is ambiguous on this point.
(k) Rebuilt from two Joint stock four-wheeled bodies on new frames.
(l) District Railway type coaches jointly owned.

There is considerable doubt concerning this Diagram list. It is incomplete, and some of the entries are open to argument. Some tidying-up has been done - for instance, the number of compartments was not included, and some of the dates were omitted. Three diagrams appear to be blank (5, 16, and 24) whether they had become obsolete or whether they were ever used at all, is a debatable point. It is certain, however, that the 1901 Joint stock was never included. Some Diagrams (e.g. 8, 9, 17, and possibly others) included second class vehicles as well as thirds, they being identical in all other particulars.

The inclusion of all three types of corridor coaches together on Diagram 37 seems out of character, since earlier entries were given different diagrams on account of very minor variations, such as an inch or two in length or wheelbase. There is also the question of the Vacuum Cleaner Van, which according to at least two authentic sources also had the diagram number 37. It could not be a goods stock diagram, since the highest number reached by goods stock was only 19. There is also a query as to whether one, or two, Vacuum Cleaner Vans were actually built.

The whole question of the LT&SR passenger stock and Diagrams is unsatisfactory, and is unlikely to be unravelled completely, especially at this juncture, 80 years after the company ceased to exist. It has been stated that apart from the three coaches scrapped after the Pitsea accident in 1897, no LT&SR coach had been withdrawn before 1912. The above statistics can only be presented as they stand, and one must make the best of them one can.

Appendix Two

Passenger Stock List

FIRST CLASS

No.	*Dia.*	*Date*	*Wheels*	*Compts*	*Length*	*Renumbered* *Mid.*	*LMS*	
1 - 8	1	1877	4	1111	26′ 0″	2609-2616		
9 - 14	1	1878	4	1111	26′ 0″	2617-2622		
15 - 16	2	1876	4	1111	26′ 0″	2635-2636		
17	3	1883	4	1111	24′ 8″	2798		
18 - 19	2	1876	4	1111	26′ 0″	2637-2638		
20 - 24	2	1883	4	1111	26′ 0″	2630-2634		
25 - 33	2	1886	4	1111	26′ 0″	2639-2647		
34 - 36	2	1886	4	1111	26′ 0″	2795-2797		
37 - 44	4	1893	6	1111	29′ 10″	2495-2500, 2602		
45 - 51	1	1877	4	1111	26′ 0″	2623-2629		
52 - 54	4	1899	6	1111	29′ 10″	2603-2606		
55 - 58	6	1901	8	1111111	48′ 0″	2475-2478	10486-10489	
59	6	1904	8	1111111	48′ 0″	2479	10490	
60	6	1904	8	111½TT½11	48′ 0″	2492	18184	
61 - 62	39	1912	8	11111111	52′ 9″	2607-2608	10540-10541	(A)
63 - 64	39	1912	8	11111111	52′ 9″	-	-	(B)
65 - 69	6	1906	8	1111111	48′ 0″	2480-2484	10491-10495	
70 - 71	6	1906	8	111½TT½11	48′ 0″	2493-2494	18185-18186	
72 - 74	6	1910	8	1111111	48′ 0″	2485-2487	10496-10498	
75 - 78	6	1911	8	1111111	48′ 0″	2488-2491	10499-10502	

COMPOSITES

1 - 12	7	1876	4	3223	24′ 9″	3855-3860		(C)
13 - 16	7	1883	4	3223	24′ 9″	3851-3854		
17 - 20	8	1891	6	31113	29′ 10″	3843-3846		
21 - 24	9	1891	6	32223	29′ 10″	3847-3850	27210	(D)
25 - 26	10	1901	8	33311333	48′ 0″	3130, 3823	17297-17298	
27	10	1904	8	33311333	48′ 0″	3824	17299	
28 - 32	10	1906	8	33311333	48′ 0″	3825-3829	17300-17304	
33 - 36	10	1908	8	33311333	48′ 0″	3830-3833	17305-17308	
37 - 41	10	1910	8	33311333	48′ 0″	3834-3838	17309-17313	
42 - 45	10	1911	8	33311333	48′ 0″	3839-3842	17314-17317	
46 - 47	37	1912	8	VT13V	47′ 6″	3126-3127	4784-4785	

THIRD CLASS

1 - 6	11	1876	4	33333	25′ 9″	2444-2448, 2469	

7 - 12	11	1876	4	33333	25′ 9″	2449-2454	26451	(E)
13 - 20	11	1877	4	33333	25′ 9″	2455-2462		
21 - 25	12	1877	4	33333	25′ 9″	2470-2474		
26 - 33	12	1877	4	33333	25′ 9″	4201-4208		
34 - 38	12	1877	4	33333	25′ 9″	4261-4265		
39 - 50	12	1880	4	33333	25′ 9″	4209-4220	26453	(F)
51	12	1883	4	33333	25′ 9″	4221		
52 - 53	13	1878	4	33333	28′ 3″	2429-2430		
54	12	1883	4	33333	25′ 9″	4222		
55, 60, 61, 66	13	1878	4	33333	28′ 3″	2431-2434		
56, 59, 62, 63	11	1877	4	33333	25′ 9″	2463-2466		
57, 58, 64, 65	12	1883	4	33333	25′ 9″	4223-4226		
67 - 100	12	1883	4	33333	25′ 9″	4227-4249		
						4251-4260	26454-26456	(G)
125 - 135	17	1891	6	333333	34′ 0″	2342-2352	26524-26531	(H)
136	15	1877	4	33333	24′ 8″	4250		
137 - 142	17	1891	6	333333	34′ 0″	2353-2358	26532-26534	(J)
143	11	1877	4	33333	25′ 9″	2467		
144 - 167	17	1891	6	333333	34′ 0″	2359-2382	26535-23555	(K)
168	17	1891	6	333333	34′ 0″	-		(L)
169 - 172	17	1891	6	333333	34′ 0″	2383-2386	26556	(M)
173 - 181	14	1886	4	33333	28′ 4½″	2418-2426		
182, 186, 187	13	1881	4	33333	28′ 3″	2435-2437		
183 - 184	14	1886	4	33333	28′ 4½″	2427-2428		
185	11	1877	4	33333	25′ 9″	2468		
188 - 218	17	1898	6	333333	34′ 0″	2387-2417	26557-26579	(P)
219 - 221	18	1901	8	33333333	46′ 0″	2259-2261	13995-13997	
222, 225, 229, 233, 234	18	1901	8	333½TT½333	46′ 0″	2324-2328	18651-18655	
223, 224, 226, 227, 228, 230, 231, 232	18	1901	8	33333333	46′ 0″	2262-2269	13998-14005	
235 - 240	18	1904	8	33333333	46′ 0″	2270-2275	14006-14011	
241 - 244	40	1912	8	3333333333	52′ 9½″	2438-2441	14295-14298	(R)
245 - 246	41	1912	8	333333333	52′ 9½″	2442-2443	14294	(S)
247 - 257								
258 - 260	18	1906	8	33333333	46′ 0″	2276-2278	14012-14014	
261	18	1906	8	333½TT½333	46′ 0″	2329	18656	
262 - 267	18	1908	8	33333333	46′ 0″	2279-2284	14015-14020	
268 - 282	18	1908	8	33333333	46′ 0″	2285-2299	14021-14035	
283 - 294	18	1910	8	33333333	46′ 0″	2300-2311	14036-14047	
295 - 298	18	1911	8	33333333	46′ 0″	2312-2315	14048-14051	
299, 306	18	1911	8	333½TT½333	46′ 0″	2330-2331	18657-18658	
300 - 305	18	1911	8	33333333	46′ 0″	2316-2321	14052-14057	

307-308	18	1911	8	33333333	46′ 0″	2322-2323	14058-14059	
309 - 318	37	1912	8	VT33V	47′ 6″	2332-2341	3066-3075	

BRAKE THIRDS

1 - 9	19	1877	4	Gd33	24′ 9″	4337-4345		(T)
10 - 13	20	1878	4	Gd33	25′ 9″	4374-4376		(U)
14 - 21	21	1883	4	3Gd3	24′ 9″	4346, 4355-4357		(V)
22 - 28	21	1884	4	3Gd3	24′ 9″	4358-4364		
29 - 32	21	1891	4	3Gd3	24′ 9″	4347-4349		
33 - 38	22	1897	6	33Gd3	34′ 0″	4318-4323	27754-27757	(W)
39 - 42	23	1899	6	3Gd3	31′ 6″	4324-4327	27774	(X)
43 - 44	21	1884	4	3Gd3	24′ 9″	3892, 4365		
45 - 50	25	1901	8	33gd33	46′ 0″	4295-4300	22845-22850	
51 - 54	21	1886	4	3Gd3	24′ 9″	4351-4354		
55 - 56	25	1904	8	33gd33	46′ 0″	4301-4302	22851	(X)
57 - 58	23	1904	6	3Gd3	31′ 6″	4328-4329	27728-27229	
59 - 60	42	1912	8	Gd3333333	52′ 9½″	4330-4331	23244-23245	(Y)
61 - 63	42	1912	8	Gd333333	52′ 9½″	4332, 4333 4368	23241-23243	(Y)
64								
65 - 67	26	1906	8	Gd3333	46′ 0″	4303-4305	22852-22854	
68 - 71	26	1908	8	Gd3333	46′ 0″	4306-4309	22855-22858	
72 - 75	26	1910	8	Gd3333	46′ 0″	4310-4313	22859-22862	
76 - 79	26	1911	8	Gd3333	46′ 0″	4314-4317	22863-22866	
80 - 83	37	1912	8	GdT3V	47′ 6″	4291-4294	6397-6400	

SALOONS

1 - 2	27	1877	4	Saloon	24′ 9″	4366-4367	
3	28	1896	6	Saloon	34′ 0″	2800	
4	43	1912	8	Directors′ Sal.	46′ 9″	2799	817

PASSENGER BRAKE VANS

1 - 7	29	1877	4	Centre van	24′ 9″	165-168/70/2/4
8 - 10	29	1883	4	Centre van	24′ 9″	175-177
11 - 17	29	1886	4	Centre van	24′ 9″	178/187/191/193/194/ 202/206.

COMPARTMENTS First, second, third class are denoted by figures, 1, 2, 3

Gd Guard's van

T Lavatory

V Vestibuled end (Corridor coaches only)

½ Half-compartment

GENERAL NOTES

(A) Reconstructed from two Joint stock bodies on new frame.
(B) As (A), but apparently not actually built, since there is no trace of Midland numbers.
(C) Only Nos. 1, 3, 6, 7, 8, 12 renumbered.
(D) Only No. 24 survived to receive an LMS number.
(E) LMS 26451 was originally No. 9.
(F) LMS 26453 was originally No. 50. Also, No. 51 was built in 1883.
(G) The three which received 1933 LMS numbers were originally Nos. 73, 91, 92.
(H) Nos. 133 and 135 were scrapped before 1932.
(J) LMS 26532-4 were originally Nos. 137, 138, 142.
(K) Nos. 149, 158, 159 were withdrawn before 1932.
(L) No. 168 was scrapped after the Pitsea accident in 1897, and not replaced. The other two Diagram 17 thirds scrapped at the same time (136, 143) were replaced in the list by old four-wheelers renumbered - though there is no trace of their original numbers.
(M) LMS 26556 was originally No. 172.
(P) Nos. 188/91/2/5, 201/10/6/7 did not receive LMS numbers.
(R) Reconstructed from two Joint stock bodies on new frame.
(S) As (R). No. 246 did not receive an LMS number.
(T) Nos. 3 and 8 were Midland 4344/4345; the others were renumbered in order.
(U) No. 11 was not renumbered in Midland stock (presumably stocked).
(V) Nos. 14, 16 and 21 did not receive Midland Numbers. May have been transferred to Midland stock, but no confirmation of this to hand.
(W) No. 33 scrapped before 1932.
(X) No. 55 scrapped before 1932.
(Y) Reconstructed from two Joint stock bodies on new frame.

Appendix Three

Whitechapel & Bow Carriages Absorbed by the Midland Railway, 1912

LT&SR No.	MR No.	Renumbered 1923	1933	Withdrawn	
61	2607		10540	7/40	First Class
62	2608		10541	6/40	First Class
241	2438		14295		Third Class
242	2439		14296		Third Class
243	2440		14297		Third Class
244	2441		14298		Third Class
245	2442		14294		Third Class
246	2443			/32	Third Class
59	4330	2731	23244	7/36	Brake Third
60	4331	3860	23245	3/55	Brake Third
61	4332	3861	23241	3/40	Brake Third
62	4333	3862	23242	7/48	Brake Third
63	4368	3895	23243	10/55	Brake Third

Plate 75: Southend Central on 3rd March, 1956 with preserved LT&SR locomotive No. 80 *Thundersley* and coach No. 283. *H.C. Casserley*

Appendix Four

Goods Stock List

1 - 100	1880	Cattle Van	Diagram 4	
101 - 200	1876	Open Goods Wagon	Diagram 1	
201 - 250	1879	Open Goods Wagon	Diagram 1	
251 - 322	1884	Ballast Wagon	Diagram 12	(A)
323 - 572	1885	Covered Goods Van	Diagram 9	(B)
573 - 582	1887	Single Bolster Wagon	Diagram 10	
583 - 682	1888	Open Goods Wagon	Diagram 1	
683 - 782	1891	Open Goods Wagon	Diagram 1	
783 - 832	1891	Covered Goods Van	Diagram 9	
833 - 882	1891	Cattle Van	Diagram 4	(C)
383 - 886	1888	Furniture Van Wagon	Diagram 15	
887 - 898	1895	Open Goods Wagon	Diagram 1	(D)
899 - 943	1896	Cattle Van	Diagram 4	(E)
944 - 1043	1897	Covered Goods Van	Diagram 9	
1044 - 1143	1898	Open Goods Wagon	Diagram 1	
1144 - 1149	1898	Furniture Van Wagon	Diagram 15	
1150 - 1199	1899	OpenGoods Wagon	Diagram 1	
1200 - 1283	1899	Covered Goods Van	Diagram 9	
1284 - 1293	1899	Single Bolster Wagon	Diagram 10	
1294 - 1299	1900	Furniture Van Wagon	Diagram 15	
1300 - 1499	1900	Open Goods Wagon	Diagram 1	
1500 - 1649	1902	Covered Goods Van	Diagram 9	(F)
1650 - 1674	1902	Ballast Wagon	Diagram 12	(G)
1675 - 1774	1904	Open Goods Wagon	Diagram 1	
1775 - 1784	1904	Double Bolster Wagon	Diagram 11	
1785 - 1786	1904	Bogie Well Wagon	Diagram 18	
1737 - 1811	1904	Gunpowder Van	Diagram 17	
1812	1907	Match Wagon	Diagram 30	(+)
1813 - 1832	1907	Single Bolster Wagon	Diagram 10	
1833 - 1852	1907	Furniture Van Wagon	Diagram 15	
1853	1883	Fire Engine Truck	Diagram 19	
1854	1905	Packing Wagon		(no diagram)
1855	1905	Crane Runner		(no diagram)
1856	1916	Locomotive Stores Wagon	Diagram 37A	(+)
1857	1911	Vacuum Cleaning Van	Diagram 37	(+)
1 - 46	1880	Goods Brake Van	Diagram 2	(H)
1 - 4	1898	Ballast Brake Van	Diagram 3	(J)

Notes

(A) Nos. 251-290 rebuilt as Open Goods Wagons, Diagram 12A, 1898.

(B) Nos. 323, 332 rebuilt as Meat Vans, Diagram 16, 1904.

(C) Nos. 874, 875 rebuilt as Meat Vans, Diagram 16, 1899.

(D) Nos. 887-898 rebuilt from Goods Brake Vans, 1895.

(E) Nos. 904, 905 rebuilt as Fish Vans, Diagram 13, 1904.
(F) Nos. 1500-1513 rebuilt as Meat Vans, Diagram 16, 1904.
(G) Purchased second-hand, 1902. Origin not recorded.
(H) Twelve rebuilt as Open Goods Wagons 1895, and four rebuilt as ballast brakes 1898.
(J) Rebuilt from Goods Brake Vans 1898.
(+) These three vehicles were on passenger Diagrams, but were numbered in goods stock, (No. 1812, 1856, 1857).

The two six-wheeled Milk Vans, Nos. 1 and 2 (in a separate series of numbers) were on goods Diagram 6, but for some extraordinary reason were also on passenger Diagram 36.
Four goods Diagrams, 5, 7, 8 and 14, appear to have been blank. The only reason which comes to mind for this, is that they must have covered some types of wagons which had become obsolete, but one cannot think what they may have been, and there are no clues in the available documents or publications.

Bibliography

The author wishes to express his most grateful thanks to the following publications which have been of great assistance in compiling this book. Also to the Real Photographs Co. and Locomotive & General Railway Photographs for their kind permission to reproduce the illustrations.

H.D. Welch *The London, Tilbury & Southend Railway* (Oakwood Press 1963)
R.J. Essery *An Illustrated History of Midland Wagons Vol. 2* (Oxford Publishing Co. 1980)
C.L. Aldrich *London Tilbury & Southend Locomotives, Past and Present* (Aldrich 1945)
E.F. Carter *Britain's Railway Liveries* (Burke Publishing Co. 1952)
George Dow *Railway Heraldry* (David & Charles 1973)
Journal of the Stephenson Locomotive Society, 1940, 1941 (Articles by K.H. Leech)
The Locomotive Magazine 1913.
The Railway Magazine; Vol. 5 (*page 1*) 1904; Vol. 28 (*page 189*) 1908: Vol. 24 (*page 283*) 1909; Vol.85 (*page 338*) 1929; Vol.97 (*page 149*) 1951.